ENVIRONMENTAL SCIENCE, ENGINEERING AND TECHNOLOGY

BETTER AIR: BENEFITS AND COSTS OF THE CLEAN AIR ACT

ENVIRONMENTAL SCIENCE, ENGINEERING AND TECHNOLOGY

Additional books in this series can be found on Nova's website
under the Series tab.

Additional E-books in this series can be found on Nova's website
under the E-books tab.

ENVIRONMENTAL SCIENCE, ENGINEERING AND TECHNOLOGY

BETTER AIR: BENEFITS AND COSTS OF THE CLEAN AIR ACT

Jessica Lincoln-Oswalt
EDITOR

Nova Science Publishers, Inc.
New York

For permission to use material from this book please contact us:
Telephone 631-231-7269; Fax 631-231-8175
Web Site: http://www.novapublishers.com

NOTICE TO THE READER

Library of Congress Cataloging-in-Publication Data

Better air : benefits and costs of the Clean Air Act / editor, Jessica Lincoln-Oswalt.
p. cm.
Includes index.
ISBN 978-1-61470-724-0 (hardcover)1. Air--Pollution--Law and legislation--United States. 2. Air quality management--United States--Cost effectiveness. 3. Environmental law--United States--Cost effectiveness. 4. United States. Clean Air Act. I. Lincoln-Oswalt, Jessica.
KF3812.B48 2011
344.7304'6342--dc23

2011026240

Published by Nova Science Publishers, Inc. + New York

CONTENTS

PREFACE

The authorities and responsibilities of the Environmental Protection Agency (EPA) derive primarily from a dozen major environmental statutes. This book provides a concise summary of one of those statutes, the Clean Air Act. It provides a brief history of federal involvement in air quality regulation and of the provisions added by legislation in 1970, 1977 and 1990. It also explains major authorities contained in the Act as well as key terms and references for more detailed information on the Act and its implementation.

Chapter 1- Section 812 of the 1990 Clean Air Act Amendments established a requirement that EPA develop periodic reports that estimate the benefits and costs of the Clean Air Act (CAA). The main goal of these reports is to provide Congress and the public with comprehensive, up-to-date, peer-reviewed information on the Clean Air Act's social benefits and costs, including improvements in human health, welfare, and ecological resources, as well as the impact of CAA provisions on the US economy. This report is the third in the Section 812 series, and is the result of EPA's Second Prospective analysis of the 1990 Amendments.

Chapter 2- This report summarizes the Clean Air Act and its major regulatory requirements. The principal statute addressing air quality concerns, the Clean Air Act was first enacted in 1955, with major revisions in 1970, 1977, and 1990. The Act requires EPA to set health-based standards for ambient air quality, sets deadlines for the achievement of those standards by state and local governments, and requires EPA to set national emission standards for large or ubiquitous sources of air pollution, including motor vehicles, power plants, and other industrial sources. In addition, the Act mandates emission controls for sources of 187 hazardous air pollutants, establishes a capand-trade program to limit acid rain, requires the prevention of significant deterioration of air quality in areas with clean air, requires a program to restore visibility impaired by regional haze in national parks and wilderness areas, and implements the Montreal Protocol to phase out most ozone-depleting chemicals.

This report describes the Act's major provisions and provides tables listing all major amendments, with the year of enactment and Public Law number, and cross-referencing sections of the Act with the major U.S. Code sections of the codified statute.

In: Better Air: Benefits and Costs of the Clean Air Act
Editor: Jessica Lincoln-Oswalt
ISBN: 978-1-61470-724-0

Chapter 1

THE BENEFITS AND COSTS OF THE CLEAN AIR ACT FROM 1990-2020

United States Environmental Protection Agency and Office of Air and Radiation

ABSTRACT

Section 812 of the 1990 Clean Air Act Amendments requires the U.S. Environmental Protection Agency to develop periodic reports that estimate the benefits and costs of the Clean Air Act. The main goal of these reports is to provide Congress and the public with comprehensive, up-to-date, peer-reviewed information on the Clean Air Act's social benefits and costs, including improvements in human health, welfare, and ecological resources, as well as the impact of the Act's provisions on the US economy. This report is the third in the Section 812 series, and is the result of EPA's Second Prospective analysis of the 1990 Amendments.

The Clean Air Act Amendments (CAAA) of 1990 augmented the significant progress made in improving the nation's air quality through the original Clean Air Act of 1970 and its 1977 amendments. The amendments built off the existing structure of the original Clean Air Act, but went beyond those requirements to tighten and clarify implementation goals and timing, increase the stringency of some federal requirements, revamp the hazardous air pollutant regulatory program, refine and streamline permitting requirements, and introduce new programs for the control of acid rain and stratospheric ozone depleters. The main purpose of this report is to document the costs and benefits of the 1990 CAAA provisions incremental to those costs and benefits achieved from implementing the original 1970 Clean Air Act and the 1977 amendments.

The analysis estimates the costs and benefits of reducing emissions of air pollutants by comparing a "*with-CAAA*" scenario that reflects expected or likely future measures implemented under the CAAA with a "*without-CAAA*" scenario that freezes the scope and stringency of emissions controls at the levels that existed prior to implementing the CAAA. There are six basic steps undertaken to complete this analysis: 1. air pollutant emissions modeling; 2. compliance cost estimation; 3. ambient air quality modeling; 4. health and environmental effects estimation; 5. economic valuation of these effects; and 6. results aggregation and uncertainty characterization.

The results of our analysis, summarized in the table below, make it abundantly clear that the benefits of the CAAA exceed its costs by a wide margin, making the CAAA a very good investment for the nation. We estimate that the annual dollar value of benefits of air quality improvements will be very large, and will grow over time as emissions control programs take their full effect, reaching a level of approximately $2.0 trillion in 2020. These benefits will be achieved as a result of CAAA-related programs and regulatory compliance actions estimated to cost approximately $65 billion in 2020. Most of these benefits (about 85 percent) are attributable to reductions in premature mortality associated with reductions in ambient particulate matter; as a result, we estimate that cleaner air will, by 2020, prevent 230,000 cases of premature mortality in that year. The remaining benefits are roughly equally divided among three categories of human health and environmental improvement: preventing premature mortality associated with ozone exposure; preventing morbidity, including acute myocardial infarctions and chronic bronchitis; and improving the quality of ecological resources and other aspects of the environment, the largest component of which is improved visibility.
The very wide margin between estimated benefits and costs, and the results of our uncertainty analysis, suggest that it is extremely unlikely that the monetized benefits of the CAAA over the 1990 to 2020 period reasonably could be less than its costs, under any alternative set of assumptions we can conceive. Our central benefits estimate exceeds costs by a factor of more than 30 to one, and the high benefits estimate exceeds costs by 90 times. Even the low benefits estimate exceeds costs by about three to one.

Estimated Monetized Benefits and Costs of the 1990 Clean Air Act Amendments.

	ANNUAL ESTIMATES			PRESENT VALUE ESTIMATE
	2000	**2010**	**2020**	**1990-2020**
Monetized Direct Compliance Costs (millions 2006$):				
Central[a]	$20,000	$53,000	$65,000	$380,000
Monetized Direct Benefits (millions 2006$):				
Low[b]	$90,000	$160,000	$250,000	$1,400,000
Central	$770,000	$1,300,000	$2,000,000	$12,000,000
High[b]	$2,300,000	$3,800,000	$5,700,000	$35,000,000
Net Benefits - Benefits minus Costs (millions 2006$):				
Low	$70,000	$110,000	$190,000	$1,000,000
Central	$750,000	$1,200,000	$1,900,000	$12,000,000
High	$2,300,000	$3,700,000	$5,600,000	$35,000,000
Benefit/Cost Ratio:				
Low[c]	5/1	3/1	4/1	4/1
Central	39/1	25/1	31/1	32/1
High[c]	115/1	72/1	88/1	92/1
Compliance Costs per Premature Mortality Avoided (2006$):				
Central	$180,000	$330,000	$280,000	Not estimated

[a] The cost estimates for this analysis are based on assumptions about future changes in factors such as consumption patterns, input costs, and technological innovation, which introduce significant uncertainty. The degree of uncertainty associated with many of the key factors, however, cannot be reliably quantified. Thus, we are unable to present specific low and high cost estimates.

[b] Low and high benefits estimates correspond to 5th and 95th percentile results from statistical uncertainty analysis, incorporating uncertainties in physical effects and valuation steps of benefits analysis.

[c] The low benefit/cost ratio reflects the ratio of the low benefits estimate to the central cost estimate, while the high ratio reflects the ratio of the high benefits estimate to the central costs estimate.

LIST OF ACRONYMS

ACS	American Cancer Society
AEO	Annual Energy Outlook (from the US Department of Energy)
AERMOD	American Meteorological Society/Regulatory Model
AIM	Architectural and Industrial Maintenance
AMI	Acute myocardial infarction
APEEP	Air Pollution Emissions Experiments and Policy analysis model
AQMS	Air Quality Modeling Subcommittee (of the Council)
AMET	Atmospheric Model Evaluation Tool
ANC	Acid Neutralizing Capacity
BenMAP	Environmental Benefits Mapping and Analysis Program
CAA	Clean Air Act of 1970
CAAA	Clean Air Act Amendments of 1990
CAIR	Clean Air Interstate Rule
CAMR	Clean Air Mercury Rule
CARB	California Air Resources Board
CAVR	Clean Air Visibility Rule
CDC	Centers for Disease Control
CGE	Computable General Equilibrium
CMAQ	Community Multi-scale Air Quality [System]
CO	Carbon monoxide
COI	Cost of illness
CONUS	Continental United States (domain in CMAQ model)
Council	Advisory Council on Clean Air Compliance Analysis
C-R	Concentration-Response
CTG	Control Techniques Guideline
CV	Contingent valuation
DDT	Dichlorodiphenyl-trichloroethane
DOE	United States Department of Energy
EC	Elemental carbon
EE	Expert elicitation
EES	Ecological Effects Subcommittee (of the Council)
EGU	Electric Generating Unit
EMPAX-CGE	Economic Model for Policy Analysis – Computable General Equilibrium
EPA	United States Environmental Protection Agency
EUS	Eastern United States (domain in CMAQ model)
EV	[Hicksian] equivalent variation
eVNA	Enhanced Voronoi Neighbor Averaging
FACA	Federal Advisory Committee Act
FASOM	Forest and Agriculture Sector Optimization Model
FRM	Federal Reference Method
GDP	Gross Domestic Product
GHG	Greenhouse gas

HAP	Hazardous Air Pollutant
HAPEM6	Hazardous Air Pollution Exposure Model, Version 6
HDDV	Heavy-Duty Diesel Vehicle
HES	Health Effects Subcommittee (of the Council)
I&M	Inspection and maintenance
IC/BC	Initial and boundary conditions
IMPROVE	Interagency Monitoring of Protected Visual Environments
IPM	Integrated Planning Model
LEV	Low-Emission Vehicle
LML	Lowest measured level
MACT	Maximum Available Control Technology
MAGIC	Model of Acidification of Groundwater in Catchments
MATS	Modeled Attainment Test Software
MCIP	Meteorology-Chemistry Interface Processor
MM5	Fifth Generation Mesoscale Model
MSA	Metropolitan statistical area
NAA	Non-Attainment Area
NAAQS	National Ambient Air Quality Standards
NAICS	North American Industry Classification System
NAPAP	National Acid Precipitation Assessment Program
NEI	National Emissions Inventory
NEMS	National Energy Modeling System
NESHAP	National Emission Standard for Hazardous Air Pollutants
NH_3	Ammonia
NH_4	Ammonium
NMMAPS	National Morbidity, Mortality, and Air Pollution Study
NO_3	Nitrate
NO_x	Nitrogen oxides
NPV	Net present value
NSPS	New Source Performance Standard
O&M	Operation and maintenance
OC	Organic carbon
OTC	Ozone Transport Commission
Pb	Lead
PCB	Polychlorinated biphenyl
PM	Particulate matter
$PM_{2.5}$	Particulate matter with an aerodynamic diameter less than 2.5 microns
PM_{10}	Particulate matter with an aerodynamic diameter less than 10 microns
PPB	Parts per billion
PRB	Powder River Basin
PSU/NCAR	Pennsylvania State University/National Center for Atmospheric Research
RACT	Reasonably Available Control Technology
RADM/RPM	Regional Acid Deposition Model/Regional Particulate Model

REMSAD	Regulatory Modeling System for Aerosols and Acid Deposition
RfC	Reference concentration
RFP	Rate of Further Progress
RIA	Regulatory Impact Analysis
RSM	Response Surface Model
RUM	Random Utility Model
SAB	Science Advisory Board
SANDWICH	Sulfates, Adjusted Nitrates, Derived Water, Inferred Carbonaceous mass, and estimated aerosol acidity (H+)) process
SCAQMD	South Coast Air Quality Management District
SIP	State Implementation Plan
SMAT	Speciated Modeled Attainment Test
SMOKE	Sparse-Matrix Operator Kernel Emissions
SO_2	Sulfur dioxide
SO_x	Sulfur oxides
SOA	Secondary organic aerosol
STN	Speciation Trends Network
SUV	Sport-Utility Vehicle
TAC	Total Annualized Cost
TSP	Total Suspended Particulates
UVb or UVB	Ultraviolet B radiation
VMT	Vehicle miles traveled
VNA	Voronoi Neighbor Averaging
VOC	Volatile organic compound
VSL	Value of statistical life
WTAC	Willingness-to-accept-compensation
WTP	Willingness-to-pay
WUS	Western United States (domain in CMAQ model)
c1gm^{-3} or c1g/m^3	Micrograms per cubic meter (unit for $PM_{2.5}$ measurement)

ACKNOWLEDGMENTS

The Project Team for the Second Prospective Study was comprised of EPA staff, and staff from a number of organizations working under contract to EPA. The project manager was Jim DeMocker, Senior Policy Analyst, EPA Office of Air and Radiation. Under EPA direction, Project Team members designed and implemented the study, and authored the study's full report, summary report, and supporting technical reports and technical memoranda. In particular, the full report and summary report of the overall Second Prospective Study were authored by Jim DeMocker of EPA and Jim Neumann of Industrial Economics, Incorporated. Major contributions to the main reports and/or key supporting reports were made by Rob Brenner and Jeneva Craig of EPA; Henry Roman, Jason Price, Maura Flight, Tyra Walsh, Lindsay Ludwig, and Nadav Tanners of Industrial Economics, Incorporated; Leland Deck of Stratus Consulting; Jim Wilson and Frank Divita of E.H.

Pechan and Associates; Sharon Douglas and Boddu Venkatesh of ICF International; Neil Wheeler of Sonoma Technologies; and Brooks Depro and Robert Beach of RTI International.

Many current and former EPA and contractor staff also made helpful contributions to the development and/or review of the study. Those who made particularly significant contributions included EPA staff Bryan Hubbell, Neal Fann, Amy Lamson, Lisa Conner, Charles Fulcher, Rich Cook, Joe Touma, Chad Bailey, Ted Palma, Norm Possiel, Brian Timin, Marc Houyoux, Larry Sorrels, Ken Davidson, and Jason Lynch; and contractor staff Andrew Bollman, Maureen Mullen and Kirstin Thesing of E.H. Pechan and Associates; Belle Hudischewskyj, Tom Myers, Yi Hua Wei, and Jay Haney of ICF International; and Martin Ross and Lauren Davis of RTI International.

During all phases of the study, from initial design to final report drafting, the Project Team and the Second Prospective Study benefitted immensely from the thoughtful, rigorous, and expert advice of the Advisory Council on Clean Air Compliance Analysis (Council) and its technical subcommittees. The Council was organized under the auspices of EPA's Science Advisory Board, which provided staff support supervised by Vanessa Vu, Director of the SAB Staff Office. The Designated Federal Official for the final Council reviews was Stephanie Sanzone of the SAB Staff Office. Other SAB Staff Office personnel who assisted in the coordination of Council reviews included Holly Stallworth, Marc Rigas, Ellen Rubin, Angela Nugent, and Anthony Maciorowski.

The Council panel providing final review of the study was chaired by Professor James K. Hammitt of Harvard University. Council members serving during the final review of this report include John Bailar (Chair of the Health Effects Subcommittee), Michelle Bell, Sylvia Brandt, Linda Bui, Dallas Burtraw, Ivan J. Fernandez (Chair of the Ecological Effects Subcommittee), Shelby Gerking, Wayne Gray, D. Alan Hansen, Nathaniel Keohane, Jonathan Levy, Richard L. Poirot, Arden Pope, Armistead (Ted) Russell (Chair of the Air Quality Modeling Subcommittee), and Michael Walsh.

In addition to the Chairs listed above, members of the technical subcommittees serving during the final review of this report included David T. Allen, David Chock, Paulette Middleton, Ralph Morris, James Price, and Chris Walcek; Elizabeth Boyer, Charles T. Driscoll, Jr., Christine Goodale, Keith G. Harrison, Allan Legge, Stephen Polasky, and Ralph Stahl, Jr.; John Fintan Hurley, Patrick Kinney, Michael T. Kleinman, Bart Ostro, and Rebecca Parkin.

In addition, valuable advice and ideas in the early stages of project design and implementation, as well as review of interim products of the study, were provided by former Council members: Trudy Ann Cameron (former Council Chair), Maureen Cropper (former Council Chair), Lauraine Chestnut, Lawrence Goulder, F. Reed Johnson, Katherine Kiel, Charles Kolstad, Nino Kuenzli, Lester B. Lave, Virginia McConnell, David Popp, and V. Kerry Smith. Former subcommittee members include: Mark Castro, Harvey E. Jeffries, Morton Lippmann, and Scott Ollinger. The Council also consulted with a number of invited experts and past panel members, including Aaron Cohen, John Evans, Christopher Frey, Dale Hattis, D. Warner North, Thomas S. Wallsten, and Ronald Wyzga.

1. INTRODUCTION

Background and Purpose

Section 812 of the 1990 Clean Air Act Amendments established a requirement that EPA develop periodic reports that estimate the benefits and costs of the Clean Air Act (CAA). The main goal of these reports is to provide Congress and the public with comprehensive, up-to-date, peer-reviewed information on the Clean Air Act's social benefits and costs, including improvements in human health, welfare, and ecological resources, as well as the impact of CAA provisions on the US economy. This report is the third in the Section 812 series, and is the result of EPA's Second Prospective analysis of the 1990 Amendments.

The first report EPA created under this authority, *The Benefits and Costs of the Clean Air Act: 1970 to 1990*, was published and conveyed to Congress in October 1997. This Retrospective analysis comprehensively assessed benefits and costs of requirements of the 1970 Clean Air Act and the 1977 Amendments, up to the passage of the Clean Air Act Amendments of 1990. The results of the Retrospective analysis showed that the nation's investment in clean air was more than justified by the substantial benefits that were gained in the form of increased health, environmental quality, and productivity. The aggregate benefits of the CAA during the 1970 to 1990 period exceeded costs by a factor of 10 to 100.

A second Section 812 report, *The Benefits and Costs of the Clean Air Act: 1990 to 2010*, was completed in November of 1999 and addressed the incremental costs and benefits of the Clean Air Act Amendments (CAAA) enacted by Congress and signed by the President in November of 1990. This First Prospective analysis addressed implementation of the CAAA over the period 1990 to 2010, and found that aggregate benefits of the Amendments alone, excluding provisions in place prior to 1990, exceeded the costs by a factor of four.

Similar to these prior analyses, this document has one primary and several secondary objectives. The main goal is to provide Congress and the public with comprehensive, up- to-date, peer-reviewed information on the CAAA's social costs and benefits, including health, welfare, and ecological benefits. Data and methods derived from the Retrospective and First Prospective analysis have already been used to assist policy- makers in refining clean air regulations over the last several years, and we hope the information continues to prove useful to Congress during future Clean Air Act reauthorizations. Beyond the statutory goals of Section 812, EPA also intends to use the results of this study to help support decisions on future investments in air pollution research. In addition, lessons learned in conducting this analysis will help better target efforts to improve the accuracy and usefulness of future prospective analyses, generated either as part of this series or as part of EPA's ongoing responsibility to estimate benefits and costs of major rulemakings.

Relationship of This Report to Other Analyses

The Clean Air Act Amendments of 1990 augmented the significant progress made in improving the nation's air quality through the original Clean Air Act of 1970 and its 1977 amendments. The amendments built off the existing structure of the original Clean Air Act, but went beyond those requirements to tighten and clarify implementation goals and timing, increase the stringency of some federal requirements, revamp the hazardous air pollutant regulatory program, refine and streamline permitting requirements, and introduce new programs for the control of acid rain and stratospheric ozone depleters. Because the 1990 Amendments represented an additional improvement to the nation's existing clean air program, the analysis summarized in this report was designed to estimate the costs and benefits of the 1990 CAAA incremental to those costs and benefits assessed in the Retrospective analysis. In economic terminology, this report addresses the marginal costs and benefits of the 1990 CAAA. Figure 1-1 below outlines this relationship among the section 812 Retrospective, the First Prospective, and the Second Prospective.

As illustrated in Figure 1-1, this report effectively updates and augments the First Prospective. This report addresses essentially the same scenario and target variables as the First Prospective, but incorporates a number of significant enhancements. First, this report extends the time period of analysis an additional ten years relative to the First Prospective, covering the period from the signing of the amendments in 1990 through 2020. Second, this report reflects updated cost and emissions estimation methods, including use of a new model suited to nonroad engine regulation and incorporation of the effects of learning-by-doing on projections of direct costs. Third, this report incorporates new information on the benefits of air pollutant regulation, including use of an integrated national-scale air quality model, more comprehensive characterization of ecological benefits, and an air toxics case study. Fourth, the report reflects investments in more comprehensive uncertainty analysis, including quantitative analyses where feasible. Finally, this report incorporates a sophisticated economy-wide model to estimate effects of the CAAA on such measures as GDP, prices, and consumer welfare. The Retrospective analysis employed a similar model for assessing the direct costs of compliance, but for the first time in this study the Agency has explored the economy- wide implications of both the direct costs *and* the health benefits of the CAAA on economic productivity, providing a much more complete picture of the full implications of CAAA regulations.

The scope of this analysis is to estimate the costs and benefits of reducing emissions of criteria pollutants under two scenarios, depicted in schematic form in Figure 1-1 below:

1. An historical, "*with-CAAA*" scenario control case that reflects expected or likely future measures implemented since 1990 to comply with rules promulgated through September 2005[1]; and
2. A counterfactual "*without CAAA*" scenario baseline case that freezes the scope and stringency of emissions controls at their 1990 levels, while allowing for changes in population and economic activity and, therefore, in emissions attributable to economic and population growth.

The Second Prospective analysis required locking in a set of emissions reductions to be used in subsequent analyses at a relatively early date (late 2005), and as a result we were compelled to forecast the implementation outcome of several pending programs. The most important of these was the then-promulgated Clean Air Interstate Rule (CAIR), which took major steps to further reduce SOx and NOx emissions from electric generating units. The rule has subsequently been vacated, and then remanded; EPA is currently considering a proposed rule to modify areas identified by the court as problematic.

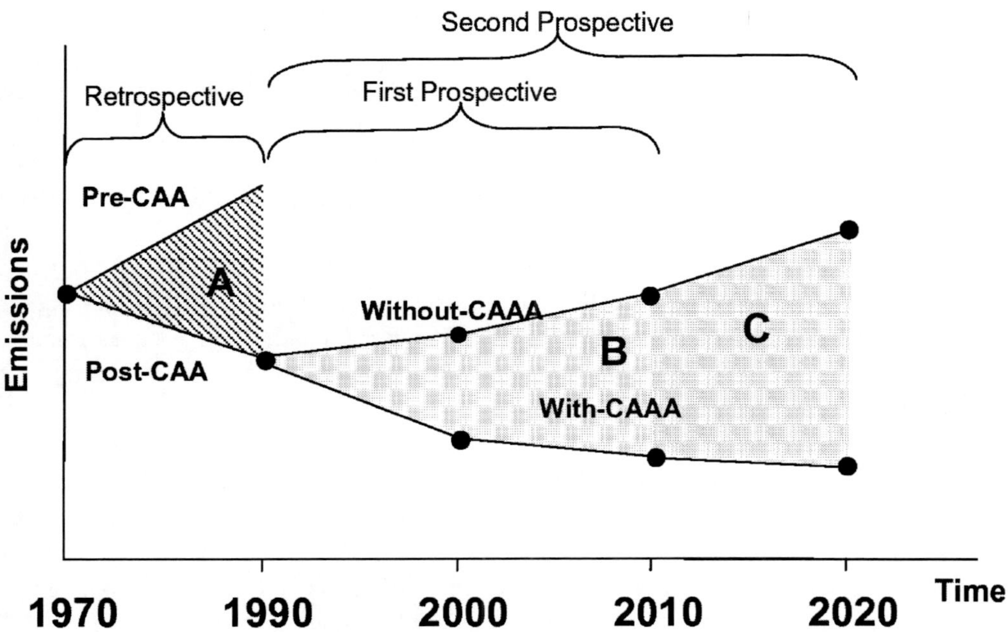

Figure 1-1. Clean Air Act Section 812 Scenarios: Conceptual Schematic.

As a result, the emissions forecasts for electric generating units incorporated in the *with-CAAA* scenario may not reflect the controls that are ultimately implemented in a modified program. We acknowledge and discuss these types of discrepancies and their impact on the outcome of our analysis in the document.

In addition, despite our efforts to comprehensively evaluate the costs and benefits of all provisions of the Clean Air Act and its Amendments, there remain a few categories of effects that are not addressed by the Retrospective or either prospective analysis. For example, this Second Prospective analysis does not assess the effect of CAAA provisions on lead exposures, primarily because the 1990 Amendments did not include major new provisions for the control of lead emissions until the NAAQS for lead was recently revisited and made significantly more stringent; the NAAQS revision was finalized after our emissions inventory development had been completed, too late for inclusion in our analysis. In addition, persistent data and model limitations preclude a full quantitative treatment of some costs and many benefits of other clean air programs. Therefore, while we considered all potentially relevant effects of the Clean Air Act and related programs, the quantitative results we present are not fully comprehensive, even for programs included in our assessment. Other, more modest omissions are acknowledged in the supporting documentation for this effort.[2]

Requirements of the 1990 Clean Air Act Amendments

This Second Prospective analysis, within the limitations discussed above, presents a comprehensive estimate of costs and benefits of the key regulatory titles of the 1990 Clean Air Act Amendments. The 1990 Amendments consist of the following eleven titles:

Title I. Establishes a detailed and graduated program for the attainment and maintenance of the National Ambient Air Quality Standards.

Title II. Regulates mobile sources and establishes requirements for reformulated gasoline and clean fuel vehicles.

Title III. Expands and modifies regulations of hazardous air pollutant emissions; and establishes a list of 189 hazardous air pollutants to be regulated.

Title IV. Establishes control programs for reducing acid rain precursors.

Title V. Requires a new permitting system for primary sources of air pollution. Title VI. Limits emissions of chemicals that deplete stratospheric ozone.

Title VII. Presents new provisions for enforcement.

Titles VIII through XI. Establish miscellaneous provisions for issues such as disadvantaged business concerns, research, training, new regulation of outer continental shelf sources, and assistance for people whose employment opportunities shift as a result of the Clean Air Act Amendments.

As part of the requirements under Title VIII, section 812 of the Clean Air Act Amendments of 1990 established a requirement that EPA analyze the costs and benefits to human health and the environment that are attributable to the Clean Air Act. In addition, section 812 directed EPA to measure the effects of this statute on economic growth, employment, productivity, cost of living, and the overall economy of the United States.

This analysis does not provide updated information on the costs and benefits of CAAA Title V regulations, which were thoroughly assessed in the First Prospective. Although Title V is believed to have yielded benefits in the efficiency of air permitting, those benefits are largely unquantified – as a result, the main effect of including Title V in the First Prospective was to increase the cost estimate by about $300 million. Similarly, we omit further consideration of Title VI regulation of the emissions of stratospheric ozone depleting substances, which was also assessed in the First Prospective. Although regulations under Title VI are continually updated and refined, the major components of Title VI were in place prior to the First Prospective and were thoroughly analyzed as part of that effort, resulting in the finding that the benefits of Title VI vastly exceeded its cost. As a result, EPA chose to focus resources in the Second Prospective on other areas and refinements. Because Titles V and VI have been previously assessed, and because Titles VII through XI are largely procedural and have mostly modest effects on air pollutant emissions and costs, this Second Prospective analysis is focused on the major emissions regulatory programs of the CAAA, which make up Titles I through IV of the statutory language.[3]

Analytical Design and Review

Target Variable

The Second Prospective analysis compares the overall health, welfare, ecological and economic benefits of the 1990 Clean Air Act Amendment programs to the costs of these programs. By examining the overall effects of the Clean Air Act, this analysis complements the Regulatory Impact Analyses (RIAs) developed by EPA over the years to evaluate individual regulations. We relied on information about the costs and benefits of specific rules provided by these RIAs, as well as other EPA analyses, in order to use resources efficiently. For this analysis, although costs can be reliably attributed to individual programs, the broad-scale approach adopted in this prospective study largely precludes reliable re-estimation of the benefits on a per-standard or per-program level. Similar to the Retrospective and First Prospective benefits analysis, this study calculates the change in incidences of adverse effects implied by changes in ambient concentrations of air pollutants. However, pollutant emissions reductions achieved contribute to changes in ambient concentrations of those, or secondarily formed, pollutants in ways that are highly complex, interactive, and often nonlinear. Although it would be possible to design specific scenarios that focused analyses only on a subset of regulations (for example, all of Title IV), those policy scenarios are not realistic. For example, exclusion of major components of the Federal rules required under the CAAA would then trigger a much greater need for reductions at the local level, in order to achieve NAAQS standards which apply at the metropolitan area scale. Further, emissions reductions achieved by the provisions of each Title, or more broadly by regulations across the CAAA provisions that apply to a specific category of emitting sources, interact with other regulations to affect the benefits implications of any emissions reduction. Therefore, benefits cannot be reliably isolated or matched to provision-specific changes in emissions or costs. Focusing on the broader target variables of overall costs and overall benefits of the Clean Air Act, the EPA Project Team adopted an approach based on construction and comparison of two distinct scenarios, briefly mentioned above: a "*without-CAAA*" and a "*with-CAAA*" scenario. The *without-CAAA* scenario essentially freezes federal, state, and local air pollution controls at the levels of stringency and effectiveness which prevailed in 1990. The *with-CAAA* scenario assumes that all federal, state, and local rules promulgated pursuant to, or in support of, the 1990 CAAA were implemented. This analysis then estimates the differences between the economic and environmental outcomes associated with these two scenarios. For more information on the specific construction of the scenarios and their relationship to historical trends, see Chapter 2 of this document.

Key Assumptions

Similar to the Retrospective and First Prospective analyses, we made two key assumptions during the scenario design process to avoid miring the analytical process in endless speculation. First, as stated above, we froze air pollution controls at 1990 levels throughout the "*without-CAAA*" scenario. Second, we assumed that the geographic distributions of population and economic activity remain the same between the two scenarios, although these distributions could be expected to change over time under both scenarios in response to differences across scenarios in income and air quality.

The first assumption is an obvious simplification. In the absence of the 1990 CAAA, one would expect to see some air pollution abatement activity, either voluntary or due to state or local regulation. It is conceivable that state and local regulation would have required air pollution abatement equal to–or even greater than–that required by the 1990 CAAA, particularly since some states, most notably California, have in the past done so. If one were to assume that state and local regulations would have been equivalent to 1990 CAAA standards, then a cost-benefit analysis of the 1990 CAAA would be a meaningless exercise since both costs and benefits would equal zero. Any attempt to predict how states' and localities' regulations would have differed from the 1990 CAAA would be too speculative to support the credibility of the ensuing analysis. Instead, the *without-CAAA* scenario has been structured to reflect the assumption that states and localities would not have invested further in air pollution control programs after 1990 in the absence of the federal CAAA. Thus, this analysis accounts for all costs and benefits of air pollution control from 1990 to 2020 and does not speculate about the fraction of costs and benefits attributable exclusively to the federal CAAA. Nevertheless, it is important to note that state and local governments and private initiatives are responsible for a significant portion of these total costs and total benefits. In the end, the benefits of air pollution controls result from partnerships among all levels of government and with the active participation and cooperation of private entities and individuals.

The second assumption concerns changing demographic patterns in response to air pollution. In the hypothetical *without-CAAA* scenario, air quality is worse than the actual 1990 conditions and the projected air quality in the *with-CAAA* scenario. It is possible that under the *without-CAAA* scenario more people, relative to the *with-CAAA* case, would move away from the most heavily polluted areas. Rather than speculate on the scale of population movement, the analysis assumes no differences in demographic patterns between the two scenarios. Similarly, the analysis assumes no differences between the two scenarios with respect to the level or spatial pattern of overall economic activity. Both scenarios do, however, reflect recent Census Bureau projections of population growth and the distribution of population across the country.

Analytic Sequence

The analysis comprises a sequence of six basic steps, summarized below and described in detail later in this report. These six steps, listed in order of completion, are:

1. emissions modeling
2. direct cost estimation
3. air quality modeling
4. health and environmental effects estimation
5. economic valuation
6. results aggregation and uncertainty characterization

Figure 1-2 summarizes the analytical sequence used to develop the prospective results; we describe the analytic process in greater detail below.

The first step of the analysis is the estimation of the effect of the 1990 CAAA on emissions sources. We generated emissions estimates through a three step process: (1) construction of an emissions inventory for the base year (1990); (2) projection of emissions for the *without-CAAA* case for three target years -- 2000, 2010, and 2020 -- assuming a freeze on emissions control regulation at 1990 levels and continued economic progress, consistent with sector-specific Department of Energy Annual Energy Outlook economic activity projections; and (3) construction of *with-CAAA* estimates for the same three target years, using the same set of economic activity projections used in the *withoutCAAA* case but with regulatory stringency, scope, and timing consistent with EPA's CAAA implementation plan (as of late 2005). The analysis reflects application of utility and other sector-specific emissions models developed and used in various offices of EPA's Office of Air and Radiation. These emissions models provide estimates of emissions of five criteria air pollutants* from each of several key emitting sectors. We provide more details in Chapter 2.

* The five pollutants are particulate matter (separate estimates for each of PM_{10} and $PM_{2.5}$), sulfur dioxide (SO_2), nitrogen oxides (NO_x), carbon monoxide (CO), and volatile organic compounds (VOCs). One of the CAA

The emissions modeling step is a critical component of the analysis, because it establishes consistency between the subsequent cost and benefit estimates that we develop. Estimates of direct compliance costs to achieve the emissions reductions estimated in the first step are generated as either an integral or subsequent output from the emissions estimation models, depending on the model used. For example, the Integrated Planning Model used to analyze the utility sector reflects a financially optimal allocation of reductions of sulfur and nitrogen oxides – taking into account the regulatory flexibility inherent in the Title IV trading programs – thereby estimating emissions reductions and compliance costs simultaneously. Direct costs are addressed in Chapter 3.

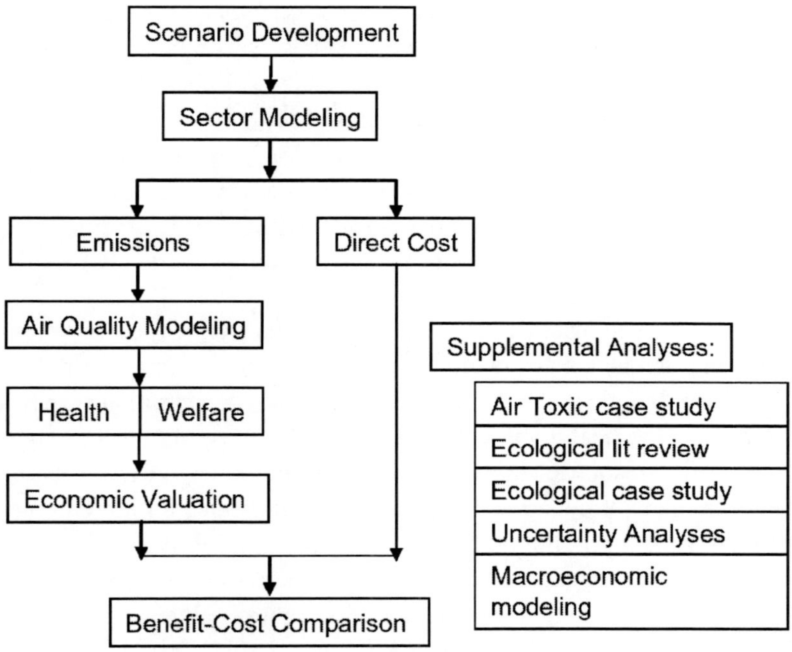

Figure 1-2. Analytic Sequence for the Second Prospective Analysis.

Emissions estimates also form the first step in estimating benefits. After the emissions inventories are developed, they are translated into estimates of air quality conditions under each scenario. For secondary particulate matter, ozone, and other air quality conditions that involve substantial non-linear formation processes and/or long-range atmospheric transport and transformation, the EPA Project Team employed EPA's Community Multi-scale Air Quality (CMAQ) system. This modeling system, for the first time in the series of Section 812 studies, provides a fully national, integrated analysis of multiple emissions and their interactions. The result is a consistent estimate of air quality for both primary and secondarily

criteria pollutants, ozone (O_3), is formed in the atmosphere through the interaction of sunlight and ozone precursor pollutants such as NOx and VOCs. We also develop estimates for ammonia (NH_3) emissions. Ammonia is not a criteria pollutant, but is an important input to the air quality modeling step because it affects secondary particulate formation. A sixth criteria pollutant, lead (Pb), is not included in this analysis since airborne emissions of lead were mostly eliminated by pre-1990 Clean Air Act programs – the recent tightening of the Pb NAAQS, necessitated by an enhanced understanding of the effects of even small exposures to airborne lead, was finalized too late to include in our scenarios. However, available estimates of the benefits and costs of the updated Pb NAAQS could be viewed as approximately additive to the results presented here.

formed pollutants, as well as deposition and visibility outcomes that represent the core of the subsequent benefit analyses. Air quality modeling is covered in Chapter 4.

Up to this point of the analysis, modeled conditions and outcomes establish the *without-CAAA* and *with-CAAA* scenarios. However, at the air quality modeling step, the analysis returns to a foundation based on actual historical conditions and data, providing a form of "ground-truthing" of the results. Specifically, actual 2000 historical air quality monitoring data are used to define the baseline conditions from which the *without-CAAA* and *with-CAAA* scenario air quality projections are constructed. We derive air quality conditions under each of the projected years of the *with-CAAA* scenario by scaling the historical data adopted for the base year (2000) by the ratio of the modeled *with-CAAA* and base year air quality. We use the same approach to estimate future year air quality for the *without-CAAA* scenario. This method takes advantage of the richness of the monitoring data on air quality, provides a realistic grounding for the benefit measures, and yet retains analytical consistency by using the same modeling process for both scenarios. The outputs of this step of the analysis are profiles for each pollutant characterizing air quality conditions at each monitoring site in the lower 48 states. This procedure also provided a means for calibrating model results in those grid cells where no monitors exist, combining model results with nearby monitor data to yield a "surface" of air quality that avoids the problems with direct extrapolation of results from monitors not located within a grid cell boundary.

The *without-CAAA* and *with-CAAA* scenario air quality profiles serve as inputs to a modeling system that translates air quality to physical outcomes (e.g., mortality, emergency room visits, or crop yield losses) through the use of concentration-response functions. Scientific literature on the health and ecological effects of air pollutants provides the source of these concentration-response functions. At this point, we derive estimates of the differences between the two scenarios in terms of incidence rates for a broad range of human health and other effects of air pollution by year, by pollutant, and by geographic area.

In the next step, we use economic valuation models or coefficients to estimate the economic value of the reduction in incidence of those adverse effects amenable to monetization. For example, a distribution of unit values derived from the economic literature provides estimates of the value of reductions in mortality risk. In addition, we compile and present benefits that cannot be expressed in economic terms. In some cases, we calculate quantitative estimates of scenario differences in the incidence of a nonmonetized effect. In many other cases, available data and techniques are insufficient to support anything more than a qualitative characterization of the change in effects. Health effects estimation and valuation are addressed in Chapter 5, and welfare effects, including ecological impacts, visibility, and agriculture and forest productivity effects, and their valuation, are addressed in Chapter 6.

Next, we compare costs and monetized benefits to provide our primary estimate of the net economic benefits of the 1990 CAAA and associated programs, and a range of estimates around that primary estimate reflecting quantified uncertainties associated with the physical effects and economic valuation steps. The monetized benefits used in the net benefit calculations reflect only a portion of the total benefits due to limitations in analytical resources, available data and models, and the state of the science. For example, in many cases we are unable to quantify or monetize the potentially large benefits of air pollution controls that result from protection of the health, structure, and function of ecosystems. In addition, although available scientific studies demonstrate clear links between air quality changes and changes in many human health effects, the available studies do not always provide the data needed to quantify and/or monetize some of these effects. Details are provided in Chapter 7.

In addition to the sequence of analyses outlined in Figure 1-2, which are focused on generating the key target variable of national net monetized benefits, a number of supplemental analyses were also conducted to provide further insights on the impacts of CAAA provisions for natural resources, health, and economic output. The first of these supplemental analyses uses the Second Prospective's national direct cost, health incidence, and health benefits valuation results to conduct further national-scale economy-wide modeling using what is known as a Computable General Equilibrium (CGE) model. The CGE model simulates, in a simplified way, shifts in markets and transactions throughout the economy that might result from CAAA provisions. It is therefore useful in assessing impacts on Gross Domestic Product (GDP), prices, and sector shifts in production (e.g., from "dirty" to "clean" industries). Most past applications of CGEs have focused on the economy-wide implications of the *costs* of complying with regulations – as a result, many prior applications, including the use of CGE in the Retrospective study, tell only half the story. Air pollution regulations not only impose direct costs, but also yield benefits, and at least some of these benefits (e.g., reduced medical expenditures, improved labor productivity owing to better health) affect market transactions in ways that can be assessed in the CGE framework. Not all benefits are amenable to analysis in a CGE, however – for example, nonmarket effects such as willingness-to-pay to avoid pain and suffering of air pollutant-linked disease cannot be incorporated. Nonetheless, this study represents one of the first broad applications of a CGE tool to regulatory costs *and* benefits. More details are provided in Chapter 8.

Two other supplemental analyses represent local-scale case studies of difficult-toquantify benefits of air pollution regulation. One is a case study of health benefits associated with air toxics control. In prior section 812 studies, benefits of air toxics programs have been largely limited to their effects on criteria pollutant outcomes. For example, many air toxics are also volatile organic compounds, and so contribute to ozone formation, an effect which can be fairly readily quantified. The direct effects of air toxics on health, however, have been more difficult to quantify, partly because of data constraints, and partly because the highly localized effects of air toxics require a level of emissions and air quality modeling resolution that is currently infeasible for a national analysis. The air toxics case study, the results of which are presented in Chapter 5, provides an example of the benefits of air toxics control for a pollutant (benzene) and geographic scope (Houston area) that is both relatively data rich and computationally manageable.

A second case study involves ecological effects, focused on the Adirondack region of New York State. This region was carefully chosen, based on the recommendation of the Advisory Council on Clean Air Compliance Analysis Ecological Effects Subcommittee (Council EES), because of its relatively high sensitivity to the effects of deposited air pollutants, because those same effects are relatively well-studied, and because methods exist to quantify and, in many cases, monetize the benefits of air pollution controls. Using the same emissions and air quality scenarios as in the overall national study, the ecological case study assesses the impact of sulfur and nitrogen deposition in the Adirondack region on aquatic resources, particularly lakes and ponds that support recreational fishing, and on commercial timber resources.

Uncertainty analyses are also conducted at each phase of the analyses. Where applicable, we present the results of a series of quantitative uncertainty analyses that test the effect of alternative methods, models, or assumptions that differ from those we used to derive the primary net benefit estimate. The primary estimate of net benefits and the range around this estimate, however, reflect our current interpretation of the available literature; our judgments regarding the best available data, models, and modeling methodologies; and the assumptions we consider most appropriate to adopt in the face of important uncertainties.

Finally, throughout the report, at the end of each chapter, we discuss the major sources of uncertainty for each analytic step. Although the impact of many of these uncertainties cannot be quantified, we qualitatively characterize the magnitude of effect on our net benefit results by assigning one of two classifications to each source of uncertainty: *potentially major* factors could, in our estimation, have effects of greater than five percent of the total net benefits; and *probably minor* factors likely have effects less than five percent of total net benefits.

The Second Prospective involved a much greater effort in uncertainty analyses than prior reports in this series. Figure 1-3 illustrates the Project Team's approach to uncertainty analysis in the Second Prospective, superimposed on the overall analytic chain for the study presented above. The grey box in Figure 1-3 represents the extent of uncertainty analysis in the first section 812 prospective analysis, which was largely limited to analysis of parameter uncertainty in the concentration-response and valuation steps of the benefits analyses. Those parameter uncertainty analyses have become standard practice in EPA analyses of air pollution program benefits, and are an integral part of the BenMAP benefits assessment tool. The results of the probabilistic modeling of these uncertainties constitute the "primary low" and "primary high" estimates presented in Table 5-7 in Chapter 5 as well as in Chapter 7.

Enhancements employed in the current analysis include both "online" analyses (shown in color), that feed information on uncertainty into the analytical chain at various points and propagate it through the remaining steps in the chain, and separate "offline" analyses and research that provide insights into the uncertainty, sensitivity, and robustness of results to alternative assumptions that are currently most easily modeled outside the main analytical process.

The online analyses consist of the selection of alternative inputs for mortality concentration-response and valuation in BenMAP, as well as an analysis of the effect on benefits of sector specific, marginal changes in PM-related emissions from the core scenarios. This online analysis substitutes EPA's Response Surface Model (RSM) for CMAQ. RSM is a less resource intensive meta-model of CMAQ used to rapidly approximate PM concentrations from alternative emissions inputs. Those analyses are described in much greater detail in the supporting uncertainty analysis report, referenced at the end of this chapter.

The bottom box in Figure 1-3 lists additional offline research and analysis we incorporated into the current study. As with the online analyses, these analyses were chosen because they address uncertainty in key analytical elements or choices that may significantly influence benefit or cost estimates. Most of these are described in this integrated report, some only briefly, but full descriptions of the data, models, and methods applied in these analyses are included in the underlying uncertainty analysis report.

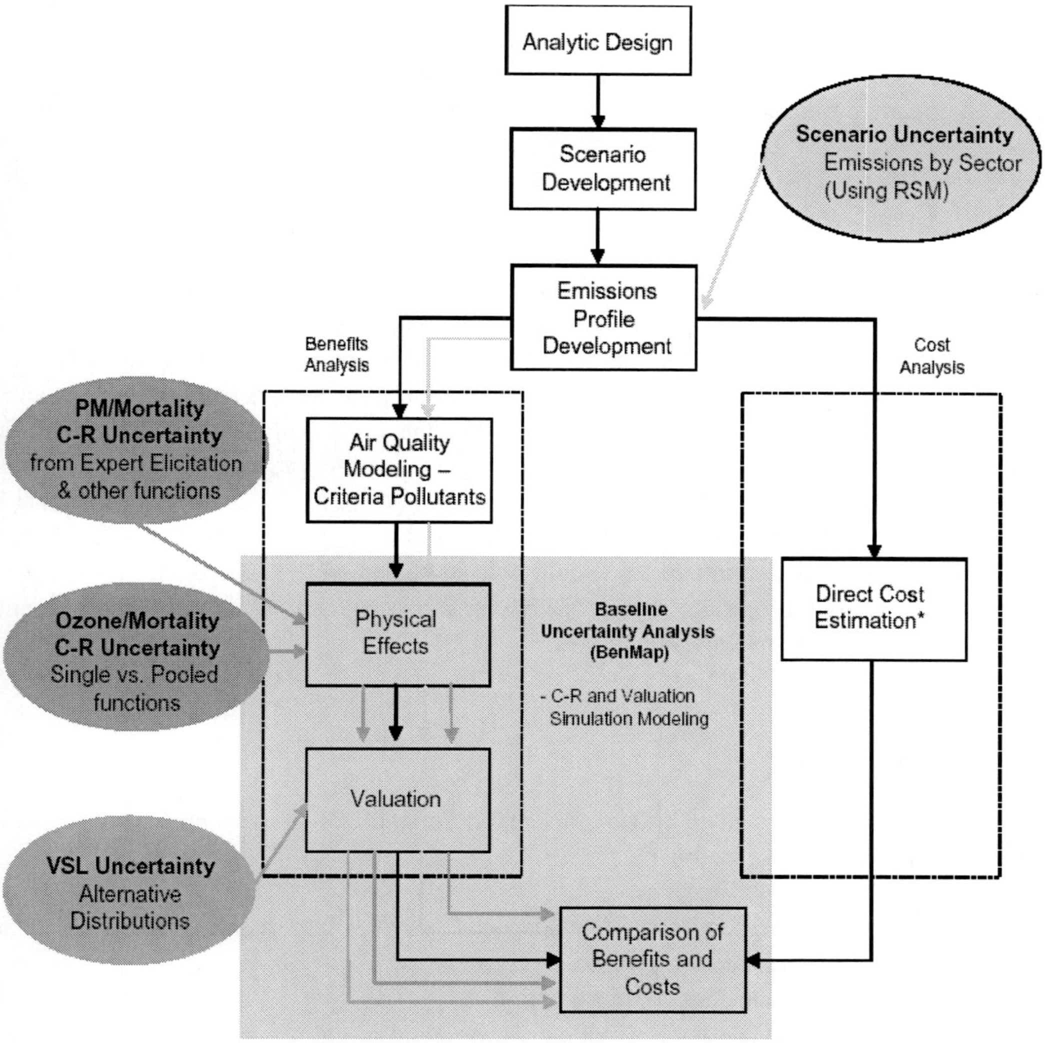

Offline Analyses

1. Dynamic versus Static Population Modeling (Benefits)
2. Cessation lag (Benefits)
3. Differential Toxicity of PM Components (Benefits)
4. Emissions and Air Quality Modeling Uncertainty Literature Review and Qualitative Analysis (Benefits)
5. Unidentified Controls (Costs)
6. Fleet Composition, I&M Failure Rates (Costs)
7. Learning Curve Assumptions (Costs)

* In addition, we perform a computable general equilibrium (CGE) analysis of costs alone and of costs and benefits, but we omit this step from the diagram because we do not conduct uncertainty analyses on the CGE modeling.

Figure 1-3. Schematic of Uncertainty Analyses.

Review Process

The 1990 CAA Amendments established a requirement that EPA consult with an outside panel of experts during the development and interpretation of the 812 studies. This panel of experts was originally organized in 1991 under the auspices of EPA's Science Advisory Board (SAB) as the Advisory Council on Clean Air Compliance Analysis (hereafter, the Council). Organizing the review committee under the SAB ensured that highly qualified experts would review the section 812 studies in an objective, rigorous, and publicly open manner consistent with the requirements and procedures of the Federal Advisory Committee Act (FACA). Council review of the present study began in 2003 with a review of the analytical design plan. Since the initial meetings, the Council and its subcommittees have met many times to review proposed data, proposed methodologies, and interim results. While the full Council retains overall review responsibility for the section 812 studies, some specific issues concerning physical effects and air quality modeling were referred to subcommittees comprised of both Council members and members of other SAB committees. The Council's Health Effects Subcommittee (HES), Air Quality Modeling Subcommittee (AQMS), and Ecological Effects Subcommittee (EES) held both in-person and teleconference meetings to review methodology proposals and modeling results and conveyed their findings and recommendations to the parent Council.

Report Organization

The remainder of the main text of this report summarizes the key methodologies and findings of our prospective study.

Chapter 2 summarizes emissions modeling and provides important additional detail on design of the regulatory scenarios.

Chapter 3 discusses the direct cost estimation.

Chapter 4 presents the air quality modeling methodology and results.

Chapter 5 describes the approaches used and principal results obtained through the human health effects estimation and valuation processes.

Chapter 6 summarizes the ecological and other welfare effects analyses, including assessments of commercial timber, agriculture, visibility, and other categories of effects.

Chapter 7 presents aggregated results of the cost and benefit estimates and describes and evaluates important uncertainties in the results.

Chapter 8 presents estimates of the effect of the Clean Air Act Amendments on economic growth, productivity, prices, household economic welfare, and the overall economy of the United States, through the application of an economy-wide economic simulation model.

Note that additional details regarding the methodologies and results of this study can be found in a series of supporting reports, available at EPA's Section 812 website (www.epa.gov/oar/sect812). These reports include the following:

Emission Projections for the Clean Air Act Second Section 812 Prospective Analysis.
Direct Cost Estimates for the Clean Air Act Second Section 812 Prospective Analysis.
Memorandum to the Files Re Documentation of Second Prospective Study Air Quality Modeling.
Health and Welfare Benefits Analyses to Support the Second Section 812 Benefit- Cost Analysis of the Clean Air Act.
Effects of Air Pollutants on Ecological Resources: Literature Review and Case Studies.
Section 812 Prospective Study of the Benefits and Costs of the Clean Air Act: Air Toxics Case Study – Health Benefits of Benzene Reductions in Houston, 1990-2020.
Uncertainty Analyses to Support the Second Section 812 Benefit-Cost Analysis of the Clean Air Act.

2. EMISSIONS

Estimation of pollutant emissions, a key component of this prospective analysis, serves as the starting point for subsequent benefit and cost estimates. We focused the emissions analysis on six major pollutants that are regulated by the Clean Air Act Amendments: volatile organic compounds (VOCs), nitrogen oxides (NO_x), sulfur dioxide (SO_2), carbon monoxide (CO), particulate matter with an aerodynamic diameter of 10 microns or less (PM_{10}), and fine particulate matter ($PM_{2.5}$). Estimates of current and future year ammonia (NH_3) emissions are also included in this study because of their importance in the atmospheric formation of fine particles in the ambient air. For each of these pollutants we projected emissions to the years 2010 and 2020 under two different scenarios:

1. An historical "*with-CAAA*" scenario control case that reflects expected or likely future measures implemented since 1990 to comply with rules promulgated through September 2005; and
2. A counterfactual "*without-CAAA*" scenario baseline case that freezes the scope and stringency of emissions controls at their 1990 levels, while allowing for changes in emissions attributable to economic and population growth.[4]

We projected emissions for five major source categories: utilities, or electricity generating units (EGUs); non-EGU industrial point sources; onroad motor vehicles; nonroad engines/vehicles; and area sources, which are smaller, more diffuse sources of pollutants that derive from many sources.[5] Table 2-1 gives examples of emissions sources for each of the five categories examined in this analysis and indicates which major pollutants are targeted by CAAA requirements in each category. The primary purpose of emissions analysis in this study is to estimate how emissions change over time and across our scenarios, so we can estimate costs of reducing emissions and the benefits of those emissions reductions for each of our target years.

This chapter consists of four sections. The first section provides an overview of our approach for developing emissions estimates. The second section summarizes our emissions projections for the years 2000, 2010, and 2020, and presents our estimates of changes in future emissions resulting from the implementation of the 1990 Amendments. The third section compares these results with estimates from the First Section 812 Prospective Analysis. Finally, we conclude this chapter with a summary of the key uncertainties associated with estimating emissions.

Overview of Approach

For four out of the five major source categories described in this report—all except electric generating units—we applied the following general method to estimate emissions:

1. Select a "base" inventory for a specific year. This involves selection of an historical year inventory from which projections will be based.
2. Select activity factors to project growth in the level of pollution-generating activity in the target years. The activity factors should provide the best possible means for representing future air pollutant emissions levels in the absence of controls.
3. Develop a database of scenario-specific emissions control factors, to represent emissions control efficiencies under the two scenarios of interest. The control factors are "layered on" to the projected emissions levels absent controls to estimate future emissions levels, taking into account those controls required for CAAA compliance .

Air pollutant emissions for the fifth category, EGUs, were estimated by application of the Integrated Planning Model (IPM), a model developed by ICF Consulting. IPM estimates EGU emissions in the 48 contiguous states and the District of Columbia through an optimization procedure that considers costs of electricity generation, costs of pollution control, and external projections of electricity demand to forecast the fuel choice, pollution control method, and generation for each unit considered in the model. We used IPM to estimate EGU emissions in both the *with-CAAA* and *without-CAAA* scenarios for 2000, 2010, and 2020.

Table 2-1. Major Emissions Source Categories.

SOURCE CATEGORY	EXAMPLES	POLLUTANTS WITH SUBSTANTIAL EMISSIONS REDUCTIONS FROM CAAA COMPLIANCE
Electricity Generating Units (EGUs)	electricity producing utilities	NO_x, SO_2
Non-EGU Industrial Point Sources	boilers, cement kilns, process heaters, turbines	NO_x, VOC, SO_2, PM_{10} $PM_{2.5}$
Onroad Motor Vehicles	buses, cars, trucks (sources that usually operate on roads and highways)	NO_x, VOC, CO
Nonroad Engines/Vehicles	aircraft, construction equipment, lawn and garden equipment, locomotives, marine engines	NO_x, VOC, CO
Area Sources	agricultural tilling, dry cleaners, open burning, wildfires	NO_x, VOC, PM_{10}, $PM_{2.5}$

Selection of Base Year Inventory

The *without-CAAA* scenario emission projections are made from a 1990 base year, while the *with-CAAA* scenario emission projections use a base year of 2000. The logic for these base year inventory choices relates to the specific definitions of the scenarios themselves. The *with-CAAA* scenario tracks compliance with CAAA requirements over time; as a result, the best basis for projecting the *with-CAAA* scenario is a current emissions inventory that incorporates decisions made since 1990 to comply with the act. The *without-CAAA* scenario, on the other hand, freezes the stringency of regulation at 1990 levels. The analysis therefore uses 1990 emission rates as a base and adjusts those emissions to account for economic activity over time. We determined that this method was less problematic than basing projections on a recent emissions inventory and trying to simulate the effect of removing CAAA emission controls currently in place. Table 2-2 summarizes the key databases that were used in this study to estimate emissions for historic years 1990 and 2000. Note that, in some cases, we determined that the best representation for year 2000 emissions was actually a later year, either 2002 or 2001. Those decisions are explained below.

For EGUs and non-EGU industrial point sources, we estimated 1990 emissions using the 1990 EPA National Emission Inventory (NEI) point source file. This file is consistent with the emission estimates used for the First Section 812 Prospective and is thought to be the most comprehensive and complete representation of point source emissions and associated activity in that year. Similarly, the 1990 EPA NEI nonpoint source file – with a few exceptions – was used to estimate 1990 area source sector emissions.[6]

For base year emissions estimates in the *with-CAAA* scenario, we drew emissions from a variety of sources. Due to resource constraints and the quality of available data, we relied on emissions estimates for years other than 2000. In the case of *with-CAAA* emissions from industrial point sources and area sources, we used the point source and nonpoint source files from the 2002 EPA NEI.[7] We chose the 2002 NEI to represent the year 2000 estimates primarily because the 2002 inventory incorporated a number of refinements in emissions estimation methods that were not included in the previous inventory, which covered 1999 emissions. We judged that the improved quality of the 2002 NEI data justified the small expected difference between emissions for these source categories in 2000 and in 2002. To estimate *with-CAAA* EGU emissions, we used data from a modified version of IPM that retrospectively modeled emissions for the year 2001.[8]

The project team estimated 1990 and 2000 emissions for the onroad and nonroad vehicle/engine sectors independently using consistent modeling approaches and activity estimates. For example, emission factors from EPA's MOBILE6.2 model were used together with data from the 1990 and 2000 NEI vehicle miles traveled (VMT) databases to estimate onroad vehicle emissions for 1990 and 2000. Similarly, EPA's NONROAD 2004 model was used to estimate 1990 and 2000 emissions for nonroad vehicles/engines.

Selection of Activity Factors for Projections

After specifying base year emissions, we projected emissions to 2000 (for the *without-CAAA* scenario), 2010, and 2020. To model emissions in the absence of controls, our general approach was to multiply an emission factor – derived from base year emissions estimates – by the level of emission-generating activity upon which the emission factor is based. These emission-generating activities vary by source category, but they are generally related to economic activity, such as transportation, energy consumption, and industrial output. Specifically, economic growth projections entered the emissions analysis in three places:

- an electricity demand forecast (included in IPM);
- a fuel consumption forecast for non-utility sectors; and
- economic growth projections that serve as activity drivers for several other sources of air pollutants.

For this analysis, we used fully integrated economic growth, energy demand, and fuel price projections to model economic growth in both the *with-CAAA* and the *withoutCAAA* scenarios. The primary advantage of this approach is that it allowed us to conduct an internally consistent analysis of economic growth across all emitting sectors. To implement this integrated approach, we chose the Department of Energy's National Energy Modeling System (NEMS), which is used to produce DOE's Annual Energy Outlook (AEO) projections. Our emissions estimates primarily rely on AEO's 2005 "reference case" scenarios. We supplemented these projections with additional forecasts from other data sources for emissions sources where we determined that AEO's energy and socioeconomic forecasts would not adequately represent growth in emissions- generating activities.[9] Table 2-3 presents the values that we used for the AEO 2005 projections for population, GDP, energy consumption, and oil price values in 2010 and 2020. For reference, the table also presents the historical values for each variable in 2002, as reported in AEO 2005. For each variable, the table shows the implied annual growth rate that AEO 2005 used to project population, GDP, energy consumption, and oil prices from 2002 to 2010 and from 2010 to 2020.[10]

Table 2-2. Base Year Emission Data Sources for the *with-* and *without-Caaa* Scenarios.

SOURCE CATEGORY	*WITHOUT-CAAA* SCENARIO – 1990	*WITH-CAAA* SCENARIO – 2000
Electricity Generating Units (EGUs)	1990 EPA Point Source NEI[1]	Estimated by the EPA Integrated Planning Model for 2001
Non-EGU Industrial Point Sources	1990 EPA Point Source NEI	2002 EPA Point Source NEI (Draft)
Onroad Motor Vehicles	MOBILE 6.2 Emission Factors and 1990 NEI VMT Database	MOBILE 6.2 Emission Factors and 2000 NEI VMT Database[2]
Nonroad Engines/Vehicles	NONROAD 2004 Model Simulation for Calendar Year 1990	NONROAD 2004 Model Simulation for Calendar Year 2000
Area Sources	1990 EPA Nonpoint Source NEI[3]	2002 EPA Nonpoint Source NEI (Final)

[1] The NEI is EPA's National Emissions Inventory, conducted every three years.

[2] The California Air Resources Board (ARB) supplied estimates for California.

[3] Adjustments were made to the 1990 nonpoint source NEI file for priority source categories.

Table 2-3. Summary of Key Driver Data Applied in Emissions Projections

VARIABLE	HISTORICAL DATA	AEO 2005 PROJECTIONS		IMPLIED ANNUAL GROWTH RATE	
	2002	2010	2020	2002-2010	2010-2020
Population (millions)	288.6	310.1	337.0	0.90%	0.83%
GDP (billion 2000 chain-weighted dollars)	$10,075	$13,084	$17,634	3.32%	3.03%
Energy Consumption (quadrillion Btu per year)	97.99	111.27	125.60	1.60%	1.22%
World Oil Price (1999 $ per barrel)	$22.17	$23.00	$26.22	0.46%	1.32%

One notable exception to the above involves the specification of $PM_{2.5}$ emissions from non-EGU point sources and area sources. After initially attempting to model $PM_{2.5}$ emissions in the *without-CAAA* scenario in 2000, 2010, and 2020 using the process described above, we determined that the resulting estimates over-attributed emissions reductions to the amendments. We applied two separate approaches to correct these emissions estimates: For emissions from area sources, we projected emissions from the two sectors responsible for the majority of emissions – construction and wood stoves – using source-specific data. For emissions from non-EGU point sources, the project team determined that emissions reductions from CAAA-mandated controls would be negligible in 2000, so we set *without-CAAA* $PM_{2.5}$ emissions equal to *with-CAAA* emissions in that year.

Applying Controls to the with-CAAA Scenario

To estimate the impact of CAAA controls on projected emissions in the *with-CAAA* scenario, we modeled the application of controls required by CAAA programs, including (among others):

- Title I VOC and NO_x reasonably available control technology (RACT) requirements in ozone nonattainment areas (NAAs);
- Title II on-road vehicle and nonroad engine/vehicle provisions;
- Title III National Emission Standards for Hazardous Air Pollutants (NESHAPs);
- Title IV programs focused on emissions from EGUs.
- Additional EGU regulations, such as the Clean Air Interstate Rule (CAIR), the Clean Air Mercury Rule (CAMR), and the Clean Air Visibility Rule (CAVR).

As a general rule, we incorporated the effects of CAAA rules promulgated through September 2005.[11] As such, we did not account for the impacts of rules promulgated after that date, such as the revised NAAQS for lead. Additionally, we modeled reductions from rules that have since been vacated, like the Clean Air Mercury Rule (CAMR) and the Clean Air Interstate Rule (CAIR), though CAIR has since been remanded. Rather than attempting to estimate the impacts of whatever rules might replace CAMR and CAIR, we modeled the rules as promulgated because that was the best information available when we made analytic commitments.

A full list of the CAAA programs modeled for each source category is presented in Table 2-4, together with the pollutants targeted by each program. For each source category, we identified factors to use in modeling the effect of emission controls required by the CAAA. For EGUs, onroad motor vehicles, and nonroad engines/vehicles, we used control factors included in the three EPA models we used to estimate base year emissions: IPM, MOBILE, and NONROAD, respectively. For non-EGU industrial point sources and area sources, we relied on control factors developed by the five Regional Planning Organizations funded by EPA to address regional air pollution issues, as well as factors developed by the California Air Resources Board.

Table 2-4. Major CAAA Programs Modeled in the *With-CAAA* Scenario

SECTOR	POLLUTANT	CAAA PROGRAMS
Electricity Generating Units (EGUs)	NO_x/ SO_2	Title IV acid rain emission allowance program; Clean Air Interstate Rule (CAIR); Clean Air Mercury Rule (CAMR); Cases and Settlements; Additional measures to meet PM and ozone NAAQS;
	NO_x	NO_x SIP Call post-2000
Non-EGU Industrial Point Sources	NO_x/VOC/ SO_2	Measures required to meet PM and ozone National Ambient Air Quality Standards (NAAQS)
	NO_x	Ozone Transport Commission (OTC) small NOx source model rule (where adopted); NOx SIP Call
	VOC	2-, 4-, 7-, and 10-year maximum achievable control technology (MACT) standards;
Onroad Motor Vehicles	NO_x/VOC/ SO_2	Tier 1 tailpipe standards (Title II); Tier 2 tailpipe standards;
	NO_x/VOC	National and California low-emission vehicle (LEV) program (Title I); Federal and California reformulated gasoline for ozone NAAQS NAAs (Title I); I/M programs for ozone and CO NAAQS NAAs (Title I); NO_x and VOC measures included in ozone NAAQS SIPs
	PM/ SO_2	Heavy-duty diesel vehicle (HDDV) standards; Diesel fuel sulfur content limits (Title II) (1993); Gasoline fuel sulfur limits; Additional measures to meet new PM NAAQS
Nonroad Engines/ Vehicles	NO_x/VOC/PM	Federal Phase I and II compression ignition (CI) and spark-ignition (S-I) engine standards; Federal commercial and recreational marine vessel standards
	NO_x/PM	Federal locomotive standards
	NO_x/PM/ SO_2	Nonroad Diesel Rule
Area Sources	NO_x/VOC/PM	RACT requirements; NOx and VOC measures included in ozone SIPs; Additional measures to meet PM and ozone NAAQS
	NO_x/VOC	Ozone Transport Commission (OTC) model rules (where adopted)
	VOC	2-, 4-, 7-, and 10-year MACT Standards; Federal VOC rules for architectural and industrial maintenance (AIM) coatings, autobody refinishing, and consumer products

Note: See Hubbell et al. (2010) for additional information regarding rules and regulations attributed to the 1990 CAAA.

Table 2-5. Emission Totals and Reductions by Pollutant – All Sectors (Thousand Tons per Year).

POLLUTANT	1990	2000			2010			2020		
		WITHOUT-CAAA	*WITH-CAAA*	REDUCTION	*WITHOUT-CAAA*	*WITH-CAAA*	REDUCTION	*WITHOUT-CAAA*	*WITH-CAAA*	REDUCTION
VOC	25,790	24,477	17,798	*6,679*	26,742	14,117	*12,626*	31,288	13,704	*17,584*
NO_x	25,917	26,688	20,837	*5,851*	28,517	13,640	*14,877*	31,740	10,092	*21,647*
CO	154,513	127,093	107,691	*19,403*	134,151	86,705	*47,447*	155,970	84,637	*71,332*
SO_2	23,143	25,129	15,319	*9,810*	26,831	10,347	*16,484*	27,912	8,272	*19,640*
PM_{10}	25,454	26,418	21,143	*5,275*	26,405	20,413	*5,992*	28,280	20,577	*7,702*
$PM_{2.5}$[1]	5,527	5,822	5,489	*333*	5,924	5,241	*682*	6,368	5,297	*1,072*
NH_3	3,656	4,136	3,983	*153*	4,405	4,224	*181*	4,787	4,587	*200*

[1] $PM_{2.5}$ without-CAAA emissions were adjusted from previously reported values by reducing emissions from non-EGU industrial point sources and area sources.

Emissions Estimation Results

Table 2-5 summarizes the national emission estimates by pollutant for each of the scenario years evaluated in this study: 2000, 2010, and 2020. As a reference, the table also presents total emissions for each pollutant in 1990. Figures 2-1 through 2-4 provide a detailed breakdown of the emissions reductions in each target year by source category for NO_x, VOC, SO_2, and primary $PM_{2.5}$. We show the breakdown of emissions reductions by source category for these pollutants because they constitute (or are precursors of) the two main air quality impacts that drive the analysis of the benefits of the CAAA: ozone and particulate matter pollution. The table and figures also incorporate our estimates of emissions reductions from local controls required to meet attainment requirements for 8-hour ozone and $PM_{2.5}$ national ambient air quality standards (NAAQS). Reductions needed for compliance, but for which we have not identified a specific pollutant reducing measure or sector to achieve the reduction, are incorporated in Table 2-5 and are presented as a separate category in Figures 2-1 through 2-4, labeled "unidentified measures."

For five of the pollutants examined—NO_x, SO_2, PM_{10}, $PM_{2.5}$, and NH_3—we estimate that emissions in the absence of the amendments would increase steadily from 1990 through 2000, 2010, and 2020, suggesting that emissions controls in place by 1990 would not be sufficient to prevent increases in pollutant emissions due to projected growth in economic activity. For the remaining two pollutants—VOC and CO—emissions decrease between 1990 and 2000 as a result of automobile tailpipe controls enacted prior to 1990, but which have delayed effects through the 1990s, before increasing from 2000 onward.

In the *with-CAAA* scenario, we estimate that emissions of SO_2 and NO_x will decrease steadily from 1990 to 2020, while emissions of VOC, CO, PM_{10}, and $PM_{2.5}$ will decrease from 1990 to 2010 before leveling off between 2010 and 2020. We also estimate that emissions of NH_3 will increase even in the presence of CAAA regulations, though at a slightly slower pace than in the *without-CAAA* scenario. NH_3 is not a specific target of CAAA regulations, but some reductions result from efforts to control other pollutants. The net result of these trends in the two scenarios is that we estimate that emissions reductions, relative to the *without-CAAA* scenario, will increase for all pollutants throughout the 2000 to 2020 period.

As Figure 2-1 shows, we estimate that reductions in NO_x emissions will increase substantially from 2000 to 2010 and from 2010 to 2020. All five major source categories contribute to these reductions in 2010 and 2020, though the largest reductions come from EGUs and on-road motor vehicles. Reductions in NO_x emissions from EGUs are driven largely by cap-and-trade programs, such as Phase II of the Ozone Transport Commission memorandum of understanding and the Clean Air Interstate Rule.[12] In the motor vehicle sector, the large reductions in NO_x emissions in 2010 and 2020 reflect both the delayed impact of Tier 1 NO_x tailpipe standards as well as the impact of Tier 2 standards, which went into effect in 2004.

Figure 2-2 shows increasing VOC emissions reductions from 2000 to 2020, with contributions from all source categories, with the exception of EGUs. The figure also shows a marked increase in on-road and nonroad emissions reductions between 2000 and 2010, reflecting both the delayed impact of Tier 1 VOC standards and the effect of low- sulfur gasoline regulations. Additionally, about half of the rules affecting nonroad sources came into effect between 2000 and 2010, explaining the increase in emissions reductions during that time. Area sources also show large emissions reductions across all three target years, driven primarily by regulations controlling evaporative emissions from solvents, though residential fireplace and woodstove emissions are also projected to decline as obsolete woodstoves are replaced with low-emitting models required by the CAAA.[13]

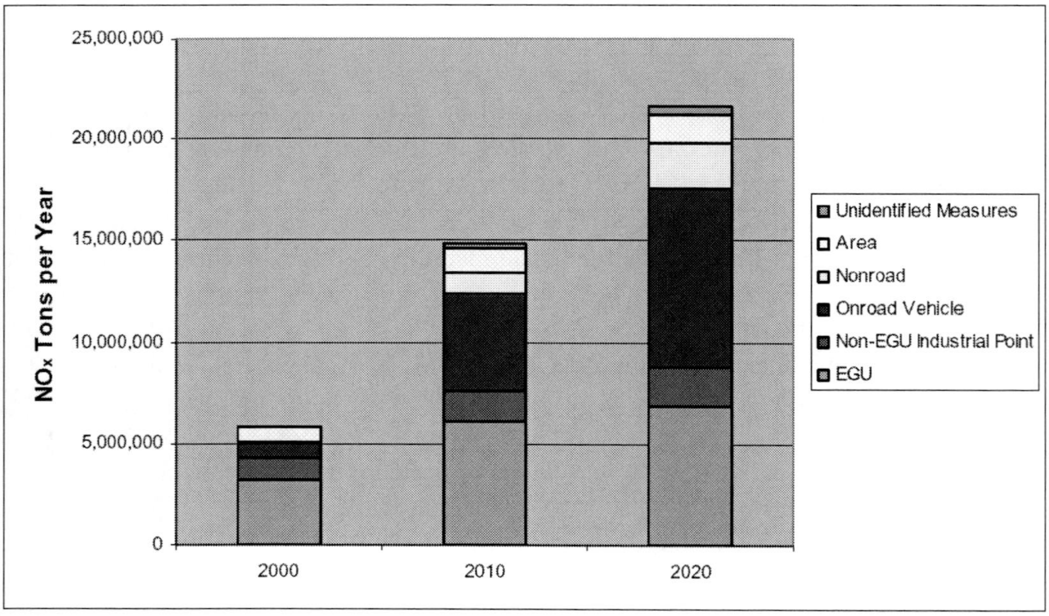

Figure 2-1. Nox Reductions Associated with CAAA Compliance by Source Category.

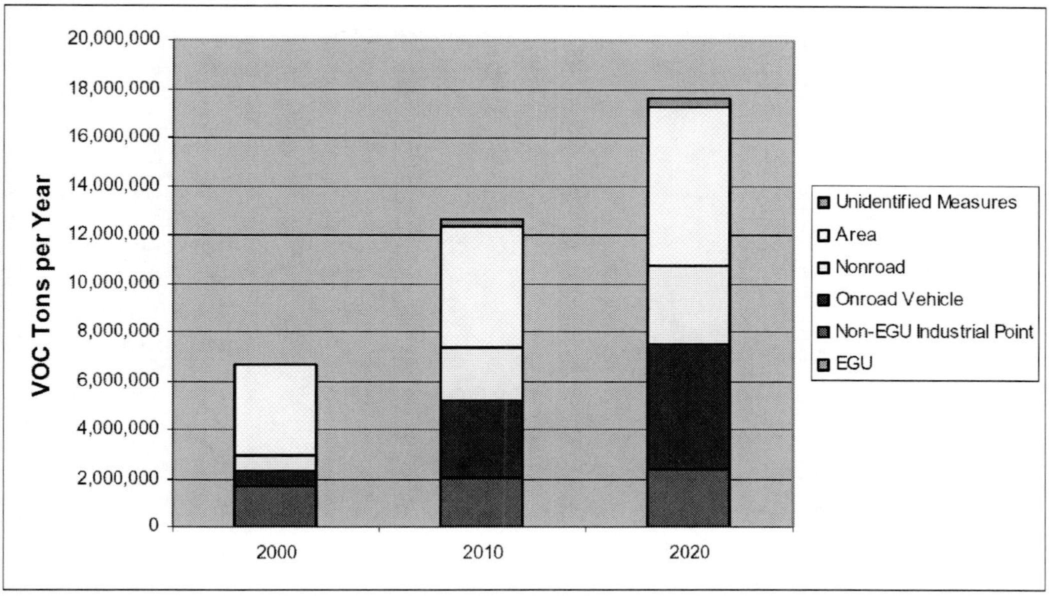

Figure 2-2. VOC Reductions Associated with CAAA Compliance by Source Category.

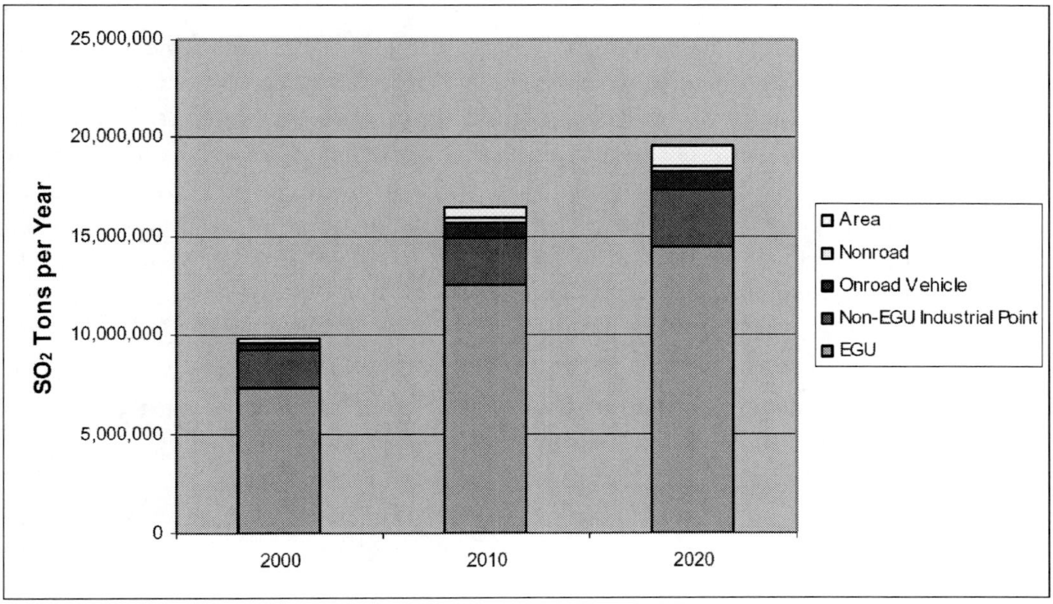

Figure 2-3. SO_2 Reductions Associated with CAAA Compliance by Source Category.

In Figure 2-3, SO_2 emissions reductions increase by more than 60 percent between 2000 and 2010, with a smaller increase between 2010 and 2020. Most reductions in SO_2 emissions in all three target years come from EGUs, with smaller contributions from nonEGU point sources and area sources as well. As with reductions in NO_x emissions, the CAIR and the Title IV cap and trade program are partly responsible for SO_2 reductions from EGUs, along with the revised $PM_{2.5}$ NAAQS.

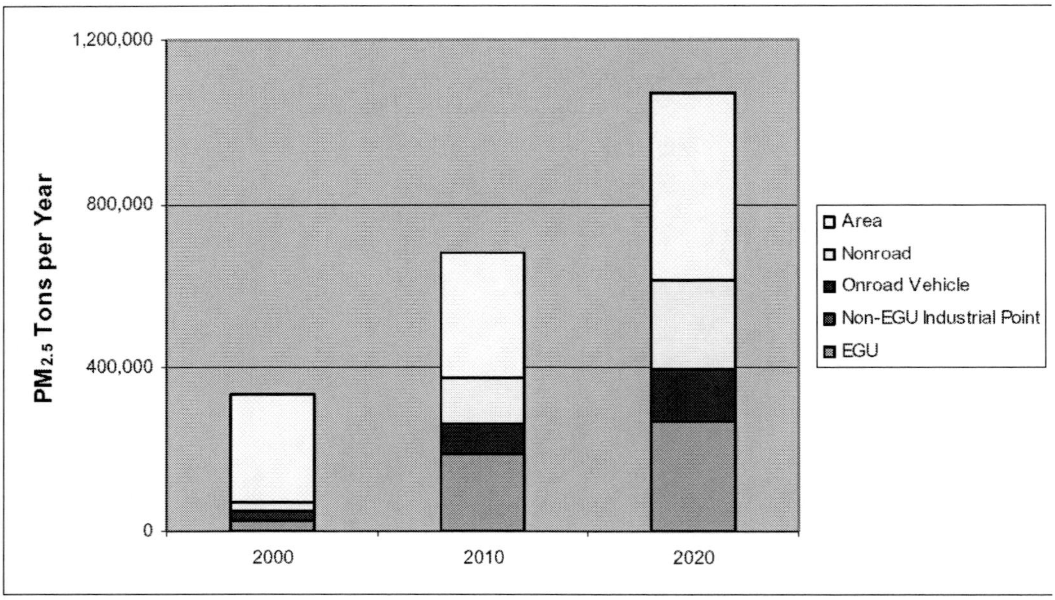

Figure 2-4. Primary $PM_{2.5}$ Reductions Associated with CAAA Compliance by Source Category.

Figure 2-4 presents reductions in $PM_{2.5}$ emissions for the three target years, with a steady increase in reductions from 2000 through 2020, as $PM_{2.5}$ NAAQS requirements ramp up. Reductions in primary fine particulate emissions are expected to come from area sources, nonroad and onroad vehicles, and EGUs. Reductions from area sources are driven largely by the replacement of obsolete residential fireplaces and wood stoves, as well as local controls on construction sites for PM NAAQS compliance. As noted above, we set $PM_{2.5}$ emissions at non-EGU industrial point sources in the *without-CAAA* scenario to be equal to emissions in the *with-CAAA* scenario, so we do not estimate that there will be any significant direct $PM_{2.5}$ emissions reductions from that source category.

Comparison of Emissions Estimates with the First Prospective Analysis

Differences in Methodology

In comparison with the First Prospective 812 Analysis, the Second Prospective includes a number of refinements and improvements in emissions estimation methods, as well as a different set of regulatory assumptions.

1. *Updated Emissions and Economic Activity Data*: Because the Second Prospective analysis was developed ten years after the First Prospective, it incorporates additional information that was not available when the First Prospective was developed. This information includes *with-CAAA* emissions estimates for the historical year 2000 as well as additional historical trend data used to project economic activity from 1990 to 2000.
2. *Additional Regulatory Requirements*: The Second Prospective Analysis accounts for several major CAA regulations that were not yet promulgated in 1996, when

decisions were made about which regulations to include in the First Prospective. These regulations include, but are not limited to, the Clean Air Interstate Rule (CAIR); the Clean Air Visibility Rule (CAVR); Tier II vehicle rules and heavy- duty diesel vehicle rules, and the local controls required for the revised 8-hour ozone and $PM_{2.5}$ NAAQS. Because of this difference, the Second Prospective Analysis models greater emissions reductions in 2000 and 2010 than were predicted in the First Prospective, as we discuss in the following section.

3. *Integrated Economic Modeling Approach*: In the First Prospective Analysis, we relied on a number of modeling tools to project future emissions, including the projections of economic activity and population growth from the Bureau of Economic Analysis, and vehicle miles traveled from EPA's MOBILE fuel consumption model. By using fully-integrated economic growth, energy demand, and fuel price projections from DOE's AEO 2005, we were able to achieve a greater degree of internal consistency in the Second Prospective Analysis.

Differences in Emissions Results

Figures 2-5 and 2-6 show estimates from the First and Second Prospective Analyses of cumulative criteria pollutant emissions and emissions reductions for 2000 and 2010, the two years that were modeled in both analyses. The figures present emissions data for the four pollutants presented in Figures 2-1 through 2-4: VOC, NO_x, SO_2, and primary $PM_{2.5}$. As Figure 2-5 shows, the Second Prospective Analysis estimates slightly higher 2000 emissions in the *without-CAAA* scenario, and slightly lower emissions in the *with- CAAA* scenario. VOC and primary $PM_{2.5}$ emissions estimates are approximately the same in both analyses, but the Second Prospective estimates reductions in combined emissions of NO_x and SO_2 of about three million tons more than in the First Prospective. As noted above, most of the difference in SO_2 emissions reductions is attributable to SO_2 controls from CAIR, but there are also substantial additional reductions attributable to reduced fuel sulfur content regulations. The difference in NO_x emissions reductions is due primarily to differences in the onroad and nonroad engine and EGU rules included in the Second Prospective, but also to corrections made in the Second Prospective to more accurately characterize the impact of the NO_x SIP Call provisions for electric generating units.

In Figure 2-6, the difference between emissions estimates in the First and Second Prospective Analyses is much more noticeable. Although the *without-CAAA* scenario emissions estimates for VOC, NOx, and SO_2 are virtually identical for the two analyses, estimates of *with-CAAA* emissions of these pollutants are all substantially lower in the Second Prospective Analysis than in the First Prospective, yielding a difference in cumulative emissions reductions of about 15 million tons. As discussed above, the Second Prospective estimates much larger emissions reductions primarily because it accounts for a number of major control programs that were not yet in place when the last analysis was published.

Uncertainty in Emissions Estimates

Table 2-6 lists several sources of uncertainty associated with generating the emissions estimates discussed in this chapter, as well as the expected direction of bias introduced by each uncertainty (if known), and the relative significance of each uncertainty in the overall

812 benefits analysis. These uncertainty sources are organized by the three factors that drive our results: identifying base-year emissions, forecasting growth in emissions-related activity, and modeling emissions controls in future years.

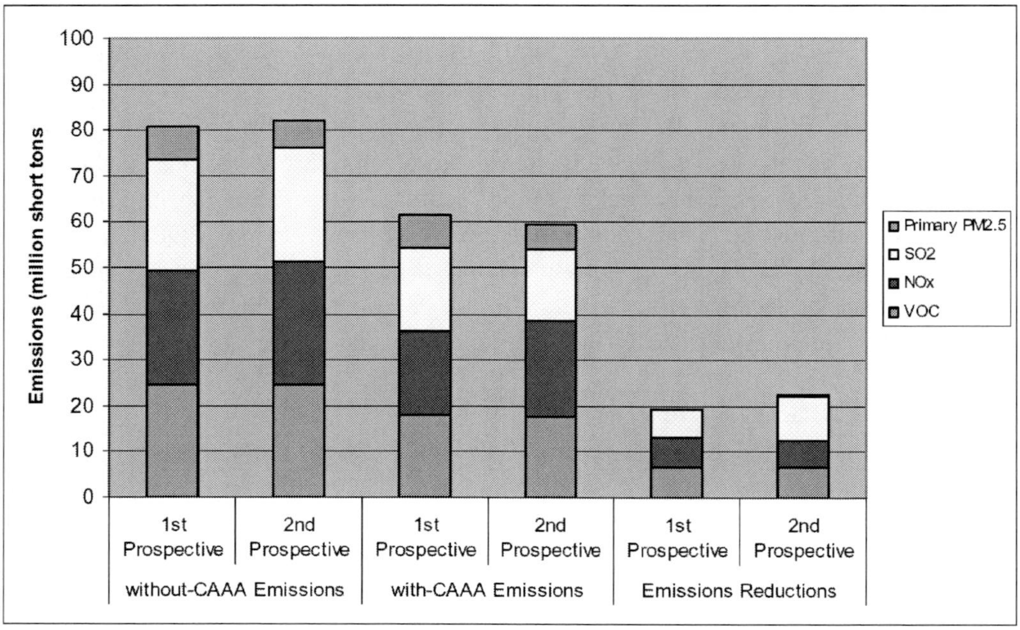

Figure 2-5. First and Second Prospective 2000 Emissions and Emissions Reductions (excluding Co and PM_{10}).

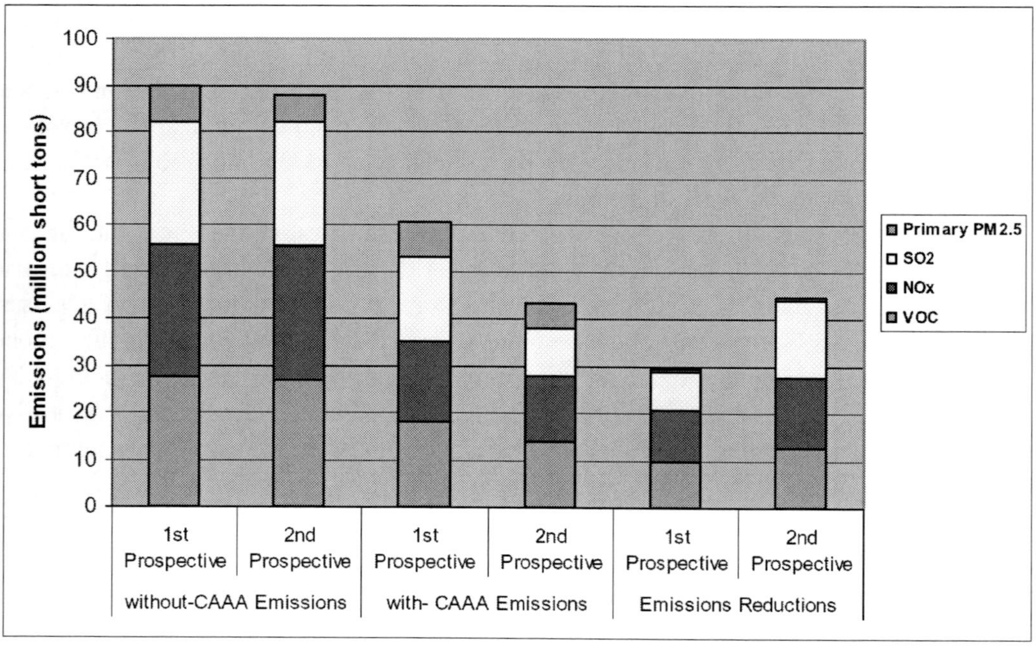

Figure 2-6 . First and Second Prospective 2010 Emissions and Emissions Reductions (excluding CO and PM_{10}).

Table 2-6. Key Uncertainties Associated with Emissions Estimation.

POTENTIAL SOURCE OF ERROR	DIRECTION OF POTENTIAL BIAS FOR NET BENEFITS	LIKELY SIGNIFICANCE RELATIVE TO KEY UNCERTAINTIES ON NET BENEFITS ESTIMATE*
Uncertainties Related To Base-Year Emissions		
Uncertainties in modeling a counterfactual emissions scenario. Estimating EGU emissions using an alternate counterfactual projection approach yielded increases in air quality impacts and health benefits of 50% relative to the core scenario's IPM-generated estimates.	Underestimate. The IPM-based counterfactual generated substantially lower benefits than the alternative counterfactual scenario specification we tested, which was based on published and readily replicated methodologies. It is possible, however, that other counterfactual specifications would yield lower benefits. It is also possible that the direction of effect might be different for other pollutant source categories where this is no accepted basis to generate an alternative counterfactual scenario estimate.	Potentially major. Analysis confirmed that IPM performs well when estimating with-CAAA emissions, but also highlighted a high degree of uncertainty in estimating counterfactual emissions. Similar uncertainties exist for emissions from other emitting sectors. There is no clear way, however, to determine what approach to estimating counterfactual emissions is superior.
Uncertainties in biogenic emissions inputs increase uncertainty in the air quality modeling estimates. Uncertainties in biogenic emissions may be large (± 80%). The biogenic inputs affect the emissions-based VOC/NOx ratio and, therefore, potentially affect the response of the modeling system to emissions changes.	Unable to determine based on current information. The biogenic emissions change overall reactivity, leading to either an underestimate or overestimate of the model's response to emission reductions.	Probably minor. Impacts for ozone and $PM_{2.5}$ results. Both oxidation potential and secondary organic aerosol formation could influence $PM_{2.5}$ formation significantly. However, biogenic emissions are assumed to be unaffected by the CAAA, so this uncertainty should not significantly affect net benefits. Furthermore, ozone benefits contribute only minimally to net benefit projections in this study.
Emissions estimated at the county level (e.g., low-level source and motor vehicle NO_x and VOC emissions) are spatially and temporally allocated based on land use, population, and	Unable to determine based on current information.	Probably minor. Potentially major for estimation of ozone, which depends largely on VOC and NO_x emissions; however, ozone benefits contribute only minimally to net benefit projections in this study.

Table 2-6. (Continued)

POTENTIAL SOURCE OF ERROR	DIRECTION OF POTENTIAL BIAS FOR NET BENEFITS	LIKELY SIGNIFICANCE RELATIVE TO KEY UNCERTAINTIES ON NET BENEFITS ESTIMATE*
other surrogate indicators of emissions activity. Uncertainty and error areintroduced to the extent that area source emissions are not perfectly spatially or temporally correlated with these indicators.		
Uncertainties Related To Growth Factors		
Economic growth factors used to project emissions are an indicator of future economic activity. These growth factors reflect uncertainty in economic forecasting as well as uncertainty in the link to emissions. IPM projections may be reasonable regionally but may introduce significant biases locally. Also, the Annual Energy Outlook 2005 growth factors do not reflect the recent economic downturn or the volatility in fuel prices since the fall of 2005.	Unable to determine based on current information.	Potentially major. The same set of growth factors are used to project emissions under both the *Without- CAAA* and *With-CAAA* scenarios, mitigating to some extent the potential for significant errors in estimating differences in emissions. Some specific locations may be more significantly influenced. We estimated gross benefits using AEO low-growth and high-growth scenarios and found differences of ±20%. However, due to nonlinearities in the benefits estimation model, we could not reliably determine in what direction over- or underestimating growth might bias net benefits estimates.
The on-road source emissions projections reflect MOBILE6.2 data on the composition of the vehicle fleet. If recent volatility in fuel prices persists or if fuel prices rise significantly (like they did in 2007 and 2008), the motor vehicle fleet may include more smaller, lower-emitting automobiles and fewer small trucks (e.g., SUVs).	Overestimate	Probably minor. Overall, fuel prices affect fleet composition at the margin, and we expect changes in fleet composition to occur gradually over long periods, suggesting that any effect would take several years to fully manifest.

Table 2-6. (Continued)

POTENTIAL SOURCE OF ERROR	DIRECTION OF POTENTIAL BIAS FOR NET BENEFITS	LIKELY SIGNIFICANCE RELATIVE TO KEY UNCERTAINTIES ON NET BENEFITS ESTIMATE*
Uncertainties Related To Emissions Control Modeling		
The *With-CAAA* scenario includes implementation of the Clean Air Mercury Rule (CAMR), which has been vacated, and Clean Air Interstate Rule (CAIR), which was vacated but has since been remanded.	Unable to determine based on current information.	Potentially major. Significance in 2020 will depend on the speed and effectiveness of implementing potential alternatives to CAIR and CAMR. In some areas, emissions reductions are expected to be overestimated, but in other areas, NO_x inhibition of ozone leads to underestimates of ozone benefits (e.g., some urban centers).
VOC emissions are dependent on evaporation, and future patterns of temperature are difficult to predict.	Underestimate. Higher temperatures in the future are more likely than lower temperatures because of climate change, and higher temperature would lead to more emissions in the *withou t-CAAA* case but controls would keep the *with- CAAA* emissions roughly constant.	Probably minor. The analysis uses meteorological data from 2002 to characterize temperatures during the 30-year period from 1990 to 2020. An acceleration of climate change (warming) could increase emissions but the increase relative to 2002 levels would not likely be significant.
Use of average temperatures (i.e., daily minimum and maximum) in estimating motor-vehicle emissions artificially reduces variability in VOC emissions.	Unable to determine based on current information.	Probably minor. Use of averages will overestimate emissions on some days and underestimate on other days. Effect is mitigated in *With-CAAA* scenarios because of more stringent evaporative controls that are in place by 2000 and 2010.
Uncertainties in the stringency, scope, timing, and effectiveness of *With-CAAA* controls included in projection scenarios.	Unable to determine based on current information.	Probably minor. Future controls could be more or less stringent, widely applicable, or effective than projected. Timing of emissions reductions may also be affected.

Table 2-6. (Continued)

POTENTIAL SOURCE OF ERROR	DIRECTION OF POTENTIAL BIAS FOR NET BENEFITS	LIKELY SIGNIFICANCE RELATIVE TO KEY UNCERTAINTIES ON NET BENEFITS ESTIMATE*
The location of the emissions reductions achieved from unidenti-fied measures is uncertain. We currently treat these reductions as if they are achieved from non-point sources, but this may not be correct in all cases.	Unable to determine based on current information.	Probably minor. Impacts from these uncertainties would be localized and would not significantly change the overall net benefit estimate.

Uncertainties Related to Base-Year Emissions

We estimated emissions from onroad motor vehicles, nonroad engines, and area sources at the county level, since these source categories are generally not tied to a specific location. Accordingly, our estimates of the spatial location of these emissions are less precise than for EGUs and industrial point sources. This uncertainty affects our ability to model changes in air quality associated with emissions reductions attributed to the CAAA. However, we expect that this uncertainty has a minor impact on the overall net benefit projections of the analysis.

A potentially major factor contributing to uncertainty in emissions estimates is our specification of the *without-CAAA* scenario. The Project Team tested the influence of an alternative scenario specification by first developing a *with-CAAA* scenario using continuous CEM data available on EPA's Clean Air Markets website.[14] Working from this scenario as a base emissions estimate for each EGU, we estimated EGU data for the *without-CAAA* scenario using an alternative counterfactual approach based on work done by Dr. A. Denny Ellerman of Massachusetts Institute of Technology.[15] The *with- CAAA* results using the alternative EGU data appear very similar to the results using the IPM EGU data, but air quality difference maps indicate that overall $PM_{2.5}$ exposures are slightly lower using the CEM data for the *with-CAAA* scenario in 2000, and $PM_{2.5}$ exposures are substantially higher using the data derived using the Ellerman counterfactual method for the *without-CAAA* scenario compared to the corresponding core scenarios.

These exposure differences carry over into benefits calculations. The health benefits of the CAAA in 2000 arrived at using the alternative EGU emissions are approximately 50 percent greater than the benefits in the 2000 core scenario. For the alternative EGU emissions scenarios, the substantial, 50 percent difference in air quality outcomes and benefits results appears to be derived from our construction of a substantially different *without-CAAA* scenario. The original motivation of the analysis was concern that the spatial pattern of emissions for the *with-CAAA* scenario for 2000 predicted by an IPM run for a historical year differed from the spatial pattern observed in the emissions monitor data for the same year. The analysis illustrated that the difference in benefits results is instead due primarily to differences in the *without-CAAA* scenario among the two alternative scenario specifications. Not surprisingly, uncertainty in estimating a counterfactual scenario is much larger than uncertainty in estimating the factual case, at least for the EGU sector.

Uncertainties Related to Growth Factors

When projecting future growth in economic activity, even the most thorough projection model must tolerate a high amount of uncertainty. The factors we used to model growth in this analysis reflect uncertainty both in the economic activity forecasted and in how this activity translates into emissions of criteria pollutants. For example, because the AEO 2005 economic growth projection predates the recent economic downturn, it is possible that we overestimate emissions in both the *with-CAAA* and *without-CAAA* scenarios. However, because we use the same growth factors to project emissions under the *with-CAAA* and *without-CAAA* scenarios, this source of uncertainty probably has a minor effect on our overall net benefits estimates. In addition, we considered projecting emissions under high-growth and low-growth AEO projection scenarios, but we did not find sufficient variation in our conclusions to justify such an analysis. For these reasons, we do not believe this is a significant factor in our results.

Similarly, our projected emissions from on-road motor vehicles are based on vehicle fleet compositions included in the MOBILE6.2 model. Any change in fuel prices that might cause a shift away from low-fuel-efficiency vehicles could cause us to overestimate emissions from this sector. However, we expect that the impact of this uncertainty on our estimate of net benefits is minor.

Uncertainties Related to Emissions Control Modeling

When modeling the *with-CAAA* scenario, we incorporated the effects of rules promulgated through September 2005. Accordingly, we did not fully account for rules promulgated since that time, such as the revised NAAQS for lead, and we modeled reductions from rules that have since been vacated, like the Clean Air Mercury Rule (CAMR) and the Clean Air Interstate Rule (CAIR), though CAIR has since been remanded. We estimated that CAMR would have only a modest impact on the pollutants we examined in this analysis, since mercury controls do not have large co-control benefits with other pollutants. However, our analysis projects that CAIR would have a large impact on NO_x and SO_2 emissions at EGUs in 2010 and 2020. Ultimately, a new rule will be promulgated to replace CAIR, and the emissions reductions, compliance costs, and locations of emissions reductions could all be different from what we modeled in this analysis. As a result, it is unclear whether our analysis overestimates or underestimates the net benefits of CAAA provisions on EGU emissions.

Estimates of emissions of volatile organic compounds are also a source of uncertainty because VOCs can be emitted through fuel combustion—like SO_2 and NO_x—as well as evaporation of volatile materials. Because evaporation rates depend largely on temperature, our estimates of future VOC emissions are influenced by the inherent difficulty of predicting future temperatures. The analysis uses projections of average daily minimum and maximum temperatures in order to predict average VOC emissions, but the resulting estimates do not adequately capture the variability of such emissions. The likely significance of this uncertainty, in terms of its impact on the overall net benefits estimated in this analysis, is probably minor.

Our future-year control assumptions are also a source of uncertainty. The flexibility allowed by the CAAA in achieving air quality standard target emission levels allows for emissions control schemes that may differ significantly from the controls modeled in this analysis. This is particularly true in the case of reductions needed for NAAQS compliance for which we have not identified a specific sector target. This analysis treats those reductions as if they come from area sources, but they could come from any of the five source categories we consider. We are not able to determine the direction of any possible bias caused by this uncertainty, but we do not expect it to have a major effect on our net benefits estimate.

3. Direct Costs

The costs of complying with the requirements of the Clean Air Act Amendments (CAAA) of 1990 will affect all levels of the U.S. economy. The impact, initially experienced through the direct costs imposed by regulations promulgated under the amendments, will also be seen in patterns of industrial production, research and development, capital investment, productivity, employment, and consumption. The purpose of the analysis summarized in this chapter is to estimate the incremental change in direct annual compliance costs from 1990 to 2020 that are attributable to the 1990 Clean Air Act Amendments.

As a measure of the direct expenditures associated with CAAA compliance, the estimates presented here represent a key stand-alone output of the Second Prospective Analysis. In addition, we use the direct cost estimates presented in this chapter to generated estimates of CAAA-related private costs that will serve as inputs in the computable general equilibrium (CGE) model used to estimate the net social costs of the CAAA on the economy as a whole.[16] Use of a CGE model allows us to estimate how compliance costs—along with expected benefits of the CAAA, such as increased labor supply—have a net impact on social welfare through interactions with labor markets and other areas of the economy. Further discussion of the CGE modeling conducted to estimate the impacts of the CAAA on net social welfare is presented in Chapter 8.

This chapter consists of four sections. The first section summarizes our approach to estimating direct compliance costs. In the second section we present the results of the cost analysis. In the third section, we discuss how cost estimates in the Second Prospective Analysis differ from those generated for the First Prospective Analysis. We conclude the chapter with a discussion of the major analytic uncertainties, including a summary of the results of quantitative sensitivity tests of key data and assumptions.

Overview of Approach

The scope of this analysis is to estimate the incremental direct costs for all criteria and hazardous air pollutant regulations issued under CAAA programs. Our approach to estimating the direct costs of CAAA compliance is closely integrated with our estimates of emissions reductions attributable to the amendments. In general, our analysis of compliance costs is driven by the results of our analysis of CAAA-related emissions reductions, and in some cases, costs and emissions reductions are measured concurrently. As with the emissions analysis presented in the previous chapter, we modeled CAAA compliance costs in 2000, 2010, and 2020 by comparing the costs of air pollution abatement in two scenarios:

- An historical "*with-CAAA*" scenario control case that reflects expected or likely future measures implemented since 1990 to comply with rules promulgated through September 2005; and
- A counterfactual "*without-CAAA*" scenario baseline case that freezes the scope and stringency of emissions controls at their 1990 levels, while allowing for changes in emissions attributable to economic and population growth.[17]

In addition, we also estimated costs separately for five major source categories: utilities, or electricity generating units (EGUs); non-EGU industrial point sources; onroad motor vehicles; nonroad engines/vehicles; and area sources. Table 2-1 gives examples of emissions sources for each of the six categories examined in this analysis. Additionally, the cost analysis considers the costs of local controls required to achieve further progress with the 8-hour Ozone NAAQS and the $PM_{2.5}$ NAAQS as a separate category. Another difference between the emissions analysis and the direct cost analysis discussed in this chapter is that, whereas the emissions analysis considered emissions of six major criteria pollutants (VOCs, NO_x,

SO_2, CO, PM_{10}, and $PM_{2.5}$) and one other pollutant which is not currently regulated under the CAAA in any form (NH_3), the cost analysis addresses CAAA provisions issued to control emissions of both criteria pollutants and hazardous air pollutants (HAPs).[18]

We estimated direct compliance costs in each source category using one of two approaches:

1. ***Cost Estimates Based on Unit Costs*** – Costs were estimated by collecting information on the costs associated with specific control measures required by CAAA regulations, or costs were calculated using estimates of the average cost per ton of pollutant emission reduced.
2. ***Cost Estimates Based on Optimization*** – Costs were estimated concurrently with emissions estimation through a cost minimizing algorithm that modeled attainment with specified emissions reduction targets. This approach was used for electric generating units, for example, where costs and emissions outcomes are outputs of the Integrated Planning Model.

Cost Estimates Based on Unit Costs

To estimate the cost of compliance CAAA regulations for most source categories, we obtained unit costs of control devices and other measures from various sources. For costs related to the 1-hour Ozone and PM_{10} National Ambient Air Quality Standards (NAAQS), we used cost data from EPA's AirControlNET database. AirControlNET links detailed data on control technologies and pollution prevention measures with EPA's National Emissions Inventory (NEI) to compute the costs associated with source- and pollutant-specific emission reductions. To calculate the cost of emissions controls on nonroad engines and vehicles, we multiplied unit cost estimates by estimates of vehicle/equipment sales and fuel consumption from the 2004 edition of EPA's NONROAD model. The NONROAD model was also used to estimate CAAA-related emissions reductions in this sector, and direct cost estimates were developed consistent with those results. For these nonroad engine and fuel rules, as well as for controls required under other parts of the CAAA, we obtained unit cost estimates from EPA's regulatory impact analyses (RIAs) as well as analyses commissioned by other organizations, such as the Ozone Transport Commission and the California Air Resources Board. Additional details on the specific data sources used to estimate unit costs for each source category are provided in the Second Prospective Cost Report.[19]

Cost Estimates Based on Optimization

We estimated control costs for EGUs using EPA's Integrated Planning Model (IPM), which determines the utility sector's least-cost strategy for meeting energy and peak demand requirements over a specified period of time, accounting for CAAA-mandated emissions caps. In the process of estimating the SO_2 and NO_x emissions that we discussed in the previous chapter, IPM also produced cost estimates for NO_x, SO_2, and mercury controls at EGUs.

We also used a least-cost optimization process to estimate the costs of local controls required to achieve further progress toward and, ultimately, approximate attainment of the 8-hour Ozone NAAQS. For each designated nonattainment area, we first modeled the application of reasonably available control technology (RACT) and inspection and maintenance (I/M) programs. Then, in areas where further emission reductions were necessary, a least-cost algorithm was used to identify and apply the control measures to meet progress and attainment requirements.[20]

Table 3-1 summarizes the cost estimation methods that we used for each source category, organized by major rules within each category.

Table 3-1. Cost Estimation Methods by Source Category and Rule (Where Applicable).

SOURCE CATEGORY	COST ESTIMATION METHOD
EGUs	IPM Least-cost optimization
Non-EGU Industrial Point Sources Ozone Transport Commission State Model Rules (NO_x/VOC): NO_x SIP Call: MACT Rules: Refinery Cases & Settlements: 1-hour Ozone NAAQS: Federal Rules (RACT, Control Technique Guidelines, National VOC Rules): Additional Measures: PM_{10} SIP Measures:	 Ozone Transport Commission -sponsored 2001 analysis AirControlNET EPA cost estimates (from 1987-1998) AirControlNET Cost/ton from 1st Prospective AirControlNET Least Cost Module SIP control cost estimates; AirControlNET
Onroad Engines and Fuels Title I NAAQS Tailpipe & Evaporative Control Standards: California and National LEV: Fuels: I/M Programs:	 EPA RIA unit costs California Air Resources Board (CARB) unit cost estimates Unit costs from First Prospective Analysis, EPA RIAs, CARB (for California standards) Costs based on information from current I/M programs
Nonroad Engines and Fuels	EPA RIA Unit Costs applied to sales and fuel consumption data provided by the NONROAD model, consistent with growth projections used to estimate emissions
Area Sources Ozone Transport Commission State Model Rules (NO_x/VOC): 1-hour Ozone NAAQS: RACT & Control Technique Guidelines: Additional Measures:	 Ozone Transport Commission-sponsored 2001 analysis Cost/ton from 1st Prospective AirControlNET
Local Controls 8-hour Ozone NAAQS: RACT & I/M: Additional (Identified) Measures: Unidentified Measures: $PM_{2.5}$ NAAQS:	 AirControlNET AirControlNET using a least-cost algorithm Assumed $15,000/ton Model SIP approach with AirControlNET unit costs

Note: Unit costs taken from earlier EPA analyses are inflated to 2006$ and adjusted to account for cost savings from learning curve impacts.
Some cost estimates for onroad and nonroad engines and fuel also reflect costs and/or savings from changes in fuel economy. These costs and savings are estimated using AEO 2005 fuel price projections.

Additional Cost Estimation Considerations

In addition to the general cost estimation methods described above, we also considered additional factors when estimating CAAA compliance costs, such as how to account for cost savings from "learning by doing," how to represent the annual costs of control measures requiring initial capital investment, and how to estimate the costs of required emissions reductions for which control measures have not yet been identified.

Learning – A significant body of literature suggests that the per unit cost of producing or using a given technology declines as experience with that technology increases over time.[21] The mechanism through which these reductions occur is not well understood, as decreases in costs may reflect several different effects, including returns to research and development, productivity spillovers from outside an industry, economies of scale, or efficiency improvements associated with increased experience with a given technology (i.e., learning-curve impacts). Given the multitude of factors that may lead to cost reductions over time, it is unclear whether such reductions should be modeled as learning-curve effects or as some other form of technological change. Nordhaus (2008) suggests that it is difficult to distinguish learning-curve effects from exogenous technological change and that learning effects, as estimated separately from technological change, will typically be overestimated. Nevertheless, the most detailed peer-reviewed empirical studies examining these cost reductions quantify a "learning rate" for different technologies and industries that represents the percentage reduction in costs associated with each doubling in the cumulative production of a technology. Based on the strength of the evidence in this literature, we incorporated the concept of the learning effect into our assessment of CAAA costs.

Where possible, we based our learning curve adjustments on learning rates presented in the empirical literature. For some sectors, however, empirical estimates of learning rates were not available. We identified learning rate estimates for SO_2 and NO_x control technologies in the EGU sector and in the onroad vehicle sector, where we used learning rates for vehicle production to estimate the impact of learning on motor vehicle engine controls. For other technologies and industries affected by the amendments, we applied a default learning rate of 10 percent, consistent with the recommendation of the Council that advised EPA on this study.[22,23]

Cost Accounting – The costs presented in this analysis are expressed as total annualized costs (TAC) in 2000, 2010, and 2020. Annualized costs include both operation and maintenance (O&M) costs and, for CAAA provisions that require investment in pollution control equipment, capital investment costs. In order to make appropriate comparisons of costs in 2000, 2010, and 2020, we annualized these investment costs over the expected life of the control equipment, rather than assigning total capital investment costs to the year in which the investment is expected to be made. We applied a discount rate of five percent to annualize capital costs over an estimated equipment life.[24] These annualized capital costs, combined with the annual O&M costs for a given pollution control measure, make up the total annualized cost estimates that we present for the three target years. Because some control measures require more capital investment than others, the degree to which our discount rate assumption affects our cost estimates varies by source category.

For CAAA-related rules that affect fuel economy, we also incorporate fuel savings or losses into our cost estimates. Where possible, we estimate the value of these benefits or costs based on fuel price projections presented in the Energy Information Administration's *Annual Energy Outlook 2005* (AEO 2005). In addition, for rules that affect the fuel economy of an engine over a period of several years, we estimate these benefits or costs as the present value of the fuel economy impacts realized over the entire life of the engine.

Local Controls for NAAQS Compliance – When estimating the costs of compliance with the 8-Hour Ozone and $PM_{2.5}$ NAAQS, we first estimated the cost of applying known and commercially available control technologies in nonattainment areas. We limited the application of these known controls to those with an estimated cost not exceeding $15,000 per ton for PM and ozone precursors (i.e., SO_2, NO_x, and VOCs). The rationale for incorporating this threshold into the analysis is that controls more costly than $15,000 per ton may not be cost effective. Thus, local air quality agencies would seek reductions from other (unidentified) control measures. This is roughly consistent with the practice of the South Coast Air Quality Management District (SCAQMD 2006) in California, which attempts to identify viable alternatives for any control requirements with an estimated cost exceeding $16,500 per ton. When costs are above this threshold, the SCAQMD also conducts more detailed cost-effectiveness and economic impact analyses of the controls.

For areas projected to remain in nonattainment with the 8-Hour Ozone NAAQS with identified controls, we estimated the costs associated with reducing emissions using additional controls not yet identified. To estimate the cost of these unidentified controls, we assumed that the cost of implementing these measures is $15,000 per ton of pollutant reduced, consistent with the cost threshold for identified controls.

Direct Compliance Cost Results

In this section we summarize the compliance cost analysis results by source category. As noted above, the control measures included in this analysis are consistent with our assumptions in the emissions analysis and reflect any post-1990 regulations promulgated (or reasonably anticipated, such as controls to meet RFP requirements) after passage of the 1990 CAAA. In general, the emissions analysis and this cost analysis reflect all of the regulations that were promulgated before September 2005. Similar to the emissions projection analysis, regulations promulgated after September 2005 (e.g., the revised Lead NAAQS) are not reflected in this report, in an effort to make the costs and benefits analyses as consistent as possible.

Table 3-2 summarizes the estimated costs of the CAAA by sector for the three analysis years: 2000, 2010 and 2020. The table shows that the direct compliance costs in 2000 are estimated to be approximately $20 billion and that these costs are dominated by the costs of motor vehicle-related provisions of the CAAA as well as MACT standards and electric utility controls. The major components of motor vehicle-related control costs in 2000 are for emission standards, fuel standards, and vehicle emission inspection programs in nonattainment areas. Motor vehicle emissions standard costs in 2000 are primarily for low emission vehicle programs, Tier 1 tailpipe standards, and on-board diagnostics. Prominent motor vehicle fuel control programs in 2000 include Federal and California reformulated gasoline. These two reformulated gasoline programs are focused primarily in serious, severe and extreme 1-hour ozone NAAQS nonattainment areas.

Because of the significant degree of uncertainty associated with estimating the costs of unidentified controls, this component of the cost analysis is reported separately in Table 3-2.

Table 3-2. Summary of 1990 CAAA Compliance Costs by Sector.

SOURCE CATEGORY	ANNUAL COST (MILLION 2006$)		
	2000	2010	2020
Electric Utilities	**$1,370**	**$6,640**	**$10,400**
Non-EGU Industrial Point Sources	**$3,130**	**$5,190**	**$5,140**
NO_x SIP Call	$0	$134	$133
MACT	$1,500	$3,010	$2,920
National VOC Rules, RACT, and New CTGs	$439	$464	$534
Refinery Settlements	$0	$295	$324
1-Hour Ozone SIP Measures	$1,030	$1,130	$1,090
PM_{10} SIP Measures	$163	$152	$146
Onroad Vehicles and Fuels	**$14,400**	**$25,700**	**$28,300**
Motor Vehicle Emission Standards	$4,400	$7,650	$7,760
California and National LEV	$562	$2,030	$2,090
Fuels	$4,820	$9,830	$11,200
Motor Vehicle I/M programs	$4,630	$6,250	$7,260
Nonroad Vehicles and Fuels	**$298**	**$359**	**$1,150**
Nonroad Engines/Vehicle Standards	$298	$219	$320
Fuels	$0	$140	$831
Area Sources	**$663**	**$693**	**$766**
RACT and New CTGs	$446	$442	$490
Ozone Transport Commission Model Rules	$134	$181	$212
1-Hour Ozone NAAQS	$82	$70	$64
Local Controls	**$0**	**$5,260**	**$6,180**
8-Hour Ozone NAAQS	$0	$4,270	$4,390
$PM_{2.5}$ NAAQS	$0	$977	$687
Clean Air Visibility Rule	$0	$0	$1,100
Sub-Total Excluding Unidentified Measures	**$19,900**	**$43,900**	**$52,000**
Additional Estimated Costs for Unidentified Controls for 8-Hour Ozone Compliance			
Non-California areas		$8,700	$8,500
California areas		$318	$5,030
TOTAL	**$19,900**	**$53,000**	**$65,500**

Note: All values are rounded to no more than three significant digits.

Table 3-2 shows that the estimated costs of complying with 1990 CAAA provisions are expected to more than double between 2000 and 2010 as areas develop and implement 8-hour ozone and $PM_{2.5}$ NAAQS State Implementation Plans (SIPs). One of the major components of CAAA compliance costs in 2010 is the estimated cost to achieve sufficient reductions of ozone precursor emissions to demonstrate 8-hour ozone NAAQS attainment. As noted above, we estimated 8-hour ozone compliance costs in two phases: first, we estimated the cost of applying known and commercially available control technologies in nonattainment areas; second, we estimated the costs associated with additional emissions reductions required to reach NAAQS attainment using controls not yet identified, at an assumed cost of $15,000 per ton. There is considerable uncertainty in this element of the cost analysis because it is unclear how individual areas will approach this issue.

The growth in costs between 2000 and 2020 partially reflects population growth during this period and the corresponding increase in emissions-generating activity (e.g., increased vehicle miles traveled). Normalized for population growth, annual costs increase from approximately $70 per capita in 2000 to $170 per capita in 2010 and $190 per capita in 2020. These results suggest that annual costs per capita grow by approximately 170 percent between 2000 and 2020, whereas annual costs (not normalized for population) grow by approximately 230 percent during this period.

Comparison of Cost Estimates with the First Prospective Analysis

In many areas, cost estimation methods in the Second Prospective Analysis were identical to those in the First Prospective, even to the point of using the same unit costs (adjusted for inflation). In general, the Second Prospective improves on the First Prospective by using more current cost estimates (where available) and more advanced least-cost optimization tools. In addition, a major methodological innovation included in the Second Prospective is the adjustment of compliance costs to account for the learning curve effects of increased experience with pollution control measures.

Figure 3-1 shows the estimated compliance costs in 2000 and 2010 from the First and Second Prospective Analyses, organized by source category. Overall, the year 2000 cost estimate presented in Table 3-2 is considerably lower than the corresponding cost estimate in the First Prospective ($27.6 billion), while the 2010 cost estimate presented in Table 3-2 is higher than the corresponding First Prospective estimate ($37.8 billion). Costs for electric utilities and area sources are significantly lower than were estimated in the First Prospective. The significant difference for utilities likely reflects differences in assumptions about the cost of obtaining low-sulfur coal from the Powder River Basin (PRB) in Wyoming. Although the Project Team was aware of the downward trend in PRB coal costs when the First Prospective was completed, this effect was not fully addressed in the data and models available at the time of the First Prospective study.

It is useful to note that the Second Prospective's $1.37 billion estimate for EGU compliance cost in 2000, which represent the pre-CAIR Title IV program requirements, fits well within the range of costs estimated in a series of *ex-post* econometric studies of compliance cost, which yield results of costs in 2000 of $1 to $1.4 billion.[25] In addition, the National Acid Precipitation Assessment Program's (NAPAP) 2005 assessment of the Clean Air Act Title IV requirements provides another basis for evaluating the reasonableness of the EGU cost estimates presented in this report (NSTC 2005). The 2005 NAPAP assessment summarizes the findings of several economic studies that estimated the cost of fully implementing the Title IV SO_2 provisions. According to NAPAP, these studies estimate annual costs ranging from $1.2 billion to $2.3 billion for full implementation in 2010, but these estimates exclude the cost of CAIR, CAMR, and some other regulations that are part of the Second Prospective estimate for 201 0.[26]

Overall, the Second Prospective cost estimates for 2010 are higher than those estimated for the First Prospective mainly because many federal motor vehicle control programs not included in the First Prospective study *with-CAAA* scenario have been promulgated since the First Prospective was completed. For the same reason, the Second Prospective cost estimates are also higher for motor vehicles in 2000, though to a lesser degree. In addition, cost estimates in the current analysis are higher than in the First Prospective because they include the costs of meeting the 8 hour ozone, $PM_{2.5}$ NAAQS and Clean Air Visibility Rule requirements in 2010. In both 2000 and 2010, estimated costs at area sources are higher in the First Prospective than in the Second Prospective, by roughly a factor of three, even though estimated emissions reductions are roughly a factor of three greater in the Second Prospective. This difference is due primarily to a much lower estimated cost per ton to reduce $PM_{2.5}$ emissions in the Second Prospective – on average, cost per ton of $PM_{2.5}$ reduced is approximately $2,000 in the Second Prospective, and was almost $20,000 in the First Prospective. One reason for the reduction is that the controls in the Second Prospective are better targeted at fine particulate control - controls in the First Prospective were actually focused on sources of PM_{10}, with $PM_{2.5}$ emissions reductions as a co-benefit. In addition, we have learned that pre-2002 NEI emissions estimates for $PM_{2.5}$ were very uncertain, suggesting that perhaps the estimated $PM_{2.5}$ emissions reductions in the First Prospective were understated.

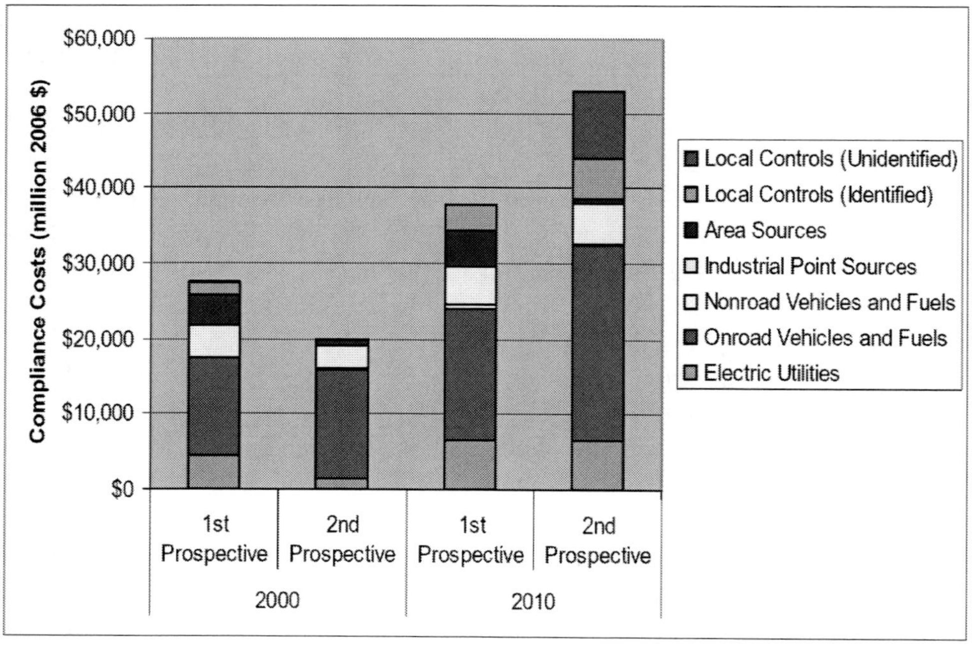

First Prospective cost estimates from U.S. EPA, The Benefits and Costs of the Clean air Act 1990 to 2010, EPA-410-R-99-001, November 1999.

Figure 3-1. First and Second Prospective Annual CAAA Compliance Costs: 2000 and 2010.

Uncertainty in Direct Cost Estimates

In a broad analysis of prospective regulatory impacts it is not possible to verify the accuracy of the full range of assumptions regarding changes in consumption patterns, input costs, and technological innovation used to estimate costs in future scenarios. Moreover, for many of the factors contributing to uncertainty, the degree or even direction of the bias is unknown or cannot be determined. Nevertheless, uncertainties and/or sensitivities can be identified and in many cases the potential measurement errors can be quantitatively characterized. In this section of the chapter, we first discuss several quantitative sensitivity analyses undertaken to characterize the impact of key assumptions on the ultimate cost analysis. The quantitative analyses presented below were chosen either because the parameter in question was a topic of discussion in the Council's review of the direct cost analysis or because we identified the parameter as potentially influential and/or uncertain. We then conclude the chapter with a qualitative discussion of the impact of both quantified and unquantified sources of uncertainty.

Quantitative Sensitivity Tests

We performed four quantitative sensitivity tests to estimate the impact of alternate assumptions on our overall cost estimates. These tests covered our assumptions regarding the cost of unidentified controls, the composition of motor vehicle sales and fleet fuel efficiency, the failure rate of I/M tests, and the default learning rate applied to sectors for which we could not identify a rate in the empirical literature. The results of these sensitivity tests on our 2020 cost estimates are presented in Table 3-3.[27]

Local Controls Analysis – Unidentified Controls

As indicated above, when estimating the cost of local controls required for further progress with the 8-hour Ozone and $PM_{2.5}$ NAAQS, we used a cost cap of $15,000 per ton to estimate the costs of identified local controls and also applied a cost of $15,000 per ton to unidentified controls. To assess the sensitivity of the local controls analysis to changes in these values, we estimated the costs of local controls based on a $10,000-perton cost cap for identified controls and a $10,000-per-ton estimated cost for unidentified controls. As indicated in Table 3-3, this alternative approach would yield lower cost estimates for both identified local controls and unidentified measures. The estimated costs of identified controls decline when the $10,000 cap is applied because controls that cost between $10,000 and $15,000 per ton are not implemented. In addition, although the application of the $10,000 cost cap increases the emissions reductions to be achieved through unidentified controls (relative to when the $15,000 cost cap is used), reducing the cost of unidentified controls to $10,000 per ton more than offsets the costs associated with these additional emissions reductions. Based on preliminary analyses conducted early in the development of the direct cost estimates, we found that in general higher thresholds do not change the emissions reductions to be achieved by unidentified controls, because few identified controls have a cost per ton higher than the $15,000 threshold used in the analysis. Accordingly, the major effect of increasing the cost cap would be to increase the estimated cost of reductions achieved by unidentified controls, whose cost is estimated based on the dollar per ton cap.

Composition of Motor Vehicle Sales and Fleet Fuel Efficiency

Our analysis of the costs associated with motor vehicle tailpipe and fuel rules is based on sales and fuel efficiency projections from the 2005 version of DOE's *Annual Energy Outlook*. Since the release of AEO 2005, however, fuel prices have been more volatile than in previous years, leading many consumers to shift to more fuel efficient vehicles, and the Department of Transportation revised the Federal Corporate Average Fuel Economy (CAFE) standards. Given these developments, AEO 2008 projects that passenger cars will make up a greater portion of light-duty vehicle sales in 2010 and 2020 than is projected by AEO 2005. AEO 2008 also assumes that the light-duty vehicle fleet will be nearly 15 percent more fuel efficient relative to the projections in AEO 2005. To assess the extent to which our cost estimates for the on-road sector would change under the alternative AEO 2008 assumptions, we estimated the cost of motor vehicle tailpipe and fuel rules for both the 2010 and 2020 target years based on the AEO 2008 data. As indicated in Table 3-3, using AEO 2008 projections increases the estimated cost of motor vehicle tailpipe standards and reduces the estimated cost of motor vehicle fuel rules in 2020. Although the alternative estimated cost of fuel rules is about 9 percent less than the primary estimate presented in Table 3-2, the reduction in estimated costs of both tailpipe and fuel CAAA motor vehicle programs in aggregate is more modest, at 3.6 percent.[28]

Vehicle Inspection Failure Rate

Our estimates of the repair costs associated with motor vehicle I&M programs employed program- and year-specific inspection failure rates derived from 2003 and 2004 data for Wisconsin I&M programs. In its June 2007 review of the Draft Direct Cost Report, the Council noted that a 2001 National Research Council report referenced a failure rate about one-seventh the value derived from the Wisconsin data.[29] To assess the sensitivity of the I&M cost analysis to the assumed failure rate for annual dynamometer-based programs, we developed alternative cost estimates for CAAA-mandated I&M programs based on the failure rate reported by the NRC. We found that the estimated cost of these programs declined by more than 40 percent when the alternative failure rates were used in place of those supporting the Second Prospective Cost Report. In addition, as indicated in Table 3-3, using these alternative values reduced total CAAA-related costs for the on- road sector by about 12 percent in 2020. This suggests that the cost estimates for the on- road sector are fairly sensitive to the assumed failure rate for I&M programs, in light of the range of failure rates obtained from readily available data sources.

Default Learning Rate

As discussed above, we adjusted total program costs to account for "learning curve" impacts (i.e., the extent to which the costs of a technology decline as experience with that technology increases over time). Wherever possible, we employed technology- or industry-specific learning rates obtained from the literature. Where industry-specific learning rates were not readily available in the empirical literature, we applied a default rate of 10 percent to the following technologies:

- Selective non-catalytic reduction at electric generating units (EGUs) (O&M costs only);
- Activated carbon injection at EGUs;
- Motor vehicle fuel rules;
- Non-road engine and fuel rules;
- Non-EGU point source controls;
- Area source controls; and
- Local controls: EGU, non-EGU point source, and area source.

We tested the sensitivity of the cost analysis to the choice of a default learning rate by re-estimating the total costs of the amendments using alternative default learning rates of 5 and 20 percent for the program areas listed above. The five percent default rate represents the low end of the range recommended by the Council, while the 20 percent value represents the central tendency presented in the peer-reviewed literature for several technologies.[30] For the sensitivity test, we did not adjust the cost estimates of program areas where the empirical literature supplied specific and applicable learning rates. As indicated in Table 3-3, the use of alternative default learning rates had only a small effect on the estimated costs of the amendments in 2020. Using a five percent default learning rate in 2020 increased the estimated cost of the amendments by 3.2 percent, while a 20 percent default learning rate reduced costs by six percent.

Table 3-3. Results of Quantitative Sensitivity Tests.

PROVISION	PRIMARY ANNUAL COST ESTIMATE FOR 2020 (BILLIONS 2006 $)	STRATEGY FOR SENSITIVITY ANALYSIS	ALTERNATIVE 2020 ESTIMATE FROM SENSITIVITY TEST (BILLIONS 2006 $)	PERCENT CHANGE FROM PRIMARY COST ESTIMATE
Local Controls (Identified and Unidentified)	$20.39	$10,000/ton cap on identified controls and $10,000/ton for unidentified controls	$16.79	-17.6%
Motor Vehicle Costs	$28.28	Use AEO 2008 projections of motor vehicle sales and fleet fuel efficiency	$27.25	-3.6%
Motor Vehicle Costs	$28.28	Use Inspection Failure Rates reported by the National Research Council	$24.82	-12.2%
Total Costs (All Source Categories)	$65.48	Use alternate default learning rate of 5 percent	$67.60	3.2%
Total Costs (All Source Categories)	$65.48	Use alternate default learning rate of 20 percent	$61.54	-6.0%

Qualitative Analysis of Key Factors Contributing to Uncertainty

In addition to the uncertainties outlined above, we identified several other areas of uncertainty related to the direct compliance costs of the amendments that we did not address quantitatively. These include the Project Team's projections of economic activity, the impact of CAAA compliance on productivity, product quality degradation resulting from the CAAA, the influence of technological innovation on CAAA compliance costs, and the impact of input substitution on the costs of complying with the amendments.

Economic Activity Projections: The cost of the amendments in 2010 and 2020 will depend in large part on the future size and composition of the U.S. economy. If the AEO 2005 economic growth projections used to estimate emissions reductions in 2010 and 2020 underestimate or overestimate economic activity, we could likewise overestimate or underestimate the costs of CAAA compliance. In addition, the particular composition of economic output in 2010 and 2020 may deviate from the AEO 2005 projections, which would also cause our cost projections to differ from the actual costs of the amendments.

Industrial Productivity: As stated in the introduction to this chapter, our cost estimates represent the direct costs of the CAAA, i.e., the expected expenditures of regulated facilities to comply with the amendments. Several peer-reviewed studies have suggested, however, that the direct costs of pollution control measures do not adequately represent the total costs of environmental protection, due to the effects of pollution abatement on industrial productivity.[31] Although our cost estimates do not capture these productivity effects, the literature is not clear on the magnitude and direction of these effects. While some studies have found that pollution control negatively affects productivity, others have found that the productivity impact is positive or ambiguous.[32]

Effects of the CAAA on Product Quality: In addition to increasing the cost of producing goods and services, CAAA requirements may also affect product quality. For example, motor vehicle emission control requirements may reduce the performance of automobiles, and changes in paint formulations (to reduce VOC emissions) may adversely affect how well paint adheres to unfinished surfaces. On the other hand, changes in product quality may also have unquantified benefits – while we capture the fuel saving benefits of many motor vehicle engine changes, the benefits of low-VOC paint in improving indoor air quality and human health are not captured in our estimates. As a result, product quality effects may reduce the welfare of households that consume products affected by the CAAA, or they may improve welfare. Households that substitute to other products due to CAAA-related quality changes (e.g., households that substitute from automobiles to light- duty trucks due to CAAA requirements that affect the performance of automobiles more than light-duty trucks) may also experience welfare losses or gains, as they would have otherwise preferred the product(s) that they would have consumed in the absence of the CAAA but may, in the balance, experience previously unrecognized gains.

Technological Innovation: The CAAA could serve as in impetus for technological innovation in the development of new, low-cost technologies or processes to reduce emissions. As indicated above, our cost estimates reflect the impact of experience-driven improvements in the productivity of existing control technologies—by accounting for learning curve impacts—but not the impact of technological innovation. Because we did not attempt to model technological innovation that might be spurred by incentives to minimize compliance costs, the Second Prospective Analysis may overestimate costs.

Input Substitution: To minimize the cost of complying with the amendments, regulated facilities may alter the mix of inputs used in the production of goods and services. With the exception of fuel switching by EGUs (as part of compliance with the Title IV Acid Rain Program and CAIR), we did not capture input substitution as a control strategy in the Second Prospective Cost Report. Ignoring the possible impact of input substitution could also cause our estimates to overstate CAAA compliance costs.

Table 3-4 lists the key sources of uncertainty noted in the quantitative and qualitative discussions above and indicates—where possible—the expected impact of the uncertainty on the net benefits estimate of the Second Prospective Analysis.

Table 3-4. Key Uncertainties Associated with Cost Estimation

POTENTIAL SOURCE OF ERROR	DIRECTION OF POTENTIAL BIAS FOR NET BENEFITS	LIKELY SIGNIFICANCE RELATIVE TO KEY UNCERTAINTIES ON NET BENEFITS ESTIMATE[1]
Uncertainty in the maximum per ton costs for local controls to comply with the 8-hour Ozone and $PM_{2.5}$ NAAQS.	Unable to determine based on current information	Probably minor. Our analysis of local controls assumes a maximum cost of $15,000 per ton for local controls implemented to comply with 8-hour Ozone and $PM_{2.5}$ NAAQS requirements.[5] Local areas may implement more costly controls to comply with the NAAQS, but technological innovation may lead to the development of less expensive controls.
Uncertainty in the projected composition of motor vehicle sales and the fuel efficiency of the motor vehicle fleet.	Unable to determine based on current information	Probably minor. We projected the composition of motor vehicle sales and the fuel efficiency of the motor vehicle fleet based on AEO 2005 data. The sensitivity analysis of alternative sales and fuel efficiency projections presented in this report suggests that this uncertainty has a small impact on net benefits.
Uncertainty regarding failure rates for motor vehicle inspections.	Unable to determine based on current information	Probably minor. The repair costs for vehicles that fail emission inspections represent a small fraction of the estimated net benefits of the amendments. The failure rate sensitivity analysis presented in this report suggests that alternative

Table 3-4. (Continued)

POTENTIAL SOURCE OF ERROR	DIRECTION OF POTENTIAL BIAS FOR NET BENEFITS	LIKELY SIGNIFICANCE RELATIVE TO KEY UNCERTAINTIES ON NET BENEFITS ESTIMATE[1]
		failure rate assumptions could have a large effect on the costs for this component of the CAAA, but only a minor effect on the estimated net benefits of the amendments as a whole.
Costs for some technologies and emissions sectors reflect default assumptions about the rates at which learning affects costs because empirical information is unavailable.	Underestimate	Probably minor. Based on the advice of the Council on Clean Air Compliance Analysis, we used a conservative learning rate of 10 percent for those sectors where no empirical data were available.[2] In contrast, the learning curve literature suggests that the average learning rate is approximately 20 percent, suggesting that learning will reduce costs more than is reflected in the present analysis.[3]
Uncertainties in the economic growth projections that form the basis of the cost analysis.	Unable to determine based on current information	Probably minor. The project team used AEO 2005 economic growth projections, which suggest that the economy will grow at an annual rate of 3.1 percent through 2025.[4] This growth rate is in line with historical GDP growth.
Incomplete characterization of certain indirect costs, such as productivity impacts for regulated industry.	Unable to determine based on current information	Probably minor. The literature on the productivity impacts of the CAAA is unclear with respect to the direction and magnitude of these effects.
Product quality degradation associated with emission control technology.	Unable to determine based on current information	Unable to determine based on current information. Conceptually, the potential for CAAA requirements to affect product quality could result in an underestimate or overestimate of the welfare effects of compliance costs, and therefore an indeterminate effect on net benefits. Unfortunately, few studies exist that address the potential product quality effects of CAAA regulations.
Exclusion of the impact of technological innovation and input substitution on compliance costs.	Underestimate	Probably minor. Minimal information is available on the potential effects of technological innovation on costs. Though input substitution is a

Table 3-4. (Continued)

POTENTIAL SOURCE OF ERROR	DIRECTION OF POTENTIAL BIAS FOR NET BENEFITS	LIKELY SIGNIFICANCE RELATIVE TO KEY UNCERTAINTIES ON NET BENEFITS ESTIMATE[1]
		potential source of cost savings, the analysis primarily models mature industries and compliance strategies which have been established as least-cost compliance paths. In addition, many regulations, such as RACT, are technology-based and may not allow for much input substitution.
Partial estimation of costs for compliance with the $PM_{2.5}$ NAAQS, due to the unavailability of emission reduction targets for non-attainment areas.	Overestimate	Probably minor. The 2006 $PM_{2.5}$ NAAQS RIA estimates that the incremental costs of residual non-attainment (i.e., costs of additional reductions from unidentified controls needed to reach attainment) are approximately $4.3 billion in 2020, yielding total cost estimates that exceed the estimates presented here by a factor of five or more.[6] However, we estimate that the costs of the $PM_{2.5}$ NAAQS represent less than 5 percent of the net benefits of the amendments.[7]
Uncertainty in the emission reduction estimates used to estimate the costs for select rules.	Unable to determine based on current information	Probably minor. Costs for many rules are not dependent on the corresponding emissions reductions (e.g., fuel sulfur limits, tailpipe standards, etc.)
Exclusion of the impact of economic incentive provisions, including banking, trading, and emissions averaging provisions.	Underestimate	Probably minor. Economic incentive provisions can substantially reduce costs, but the major economic programs for trading of sulfur and nitrogen dioxide emissions are reflected in the analysis.
Potential for overestimation biases in engineering cost estimates.	Underestimate	Probably minor. A study by Harrington, Morgenstern, and Nelson (1999) evaluated the accuracy of EPA and OSHA estimates of 25 *exante* regulatory cost estimates relative to *ex post* studies of actual costs, and concluded that initial cost estimates by EPA tend to overstate costs. The source of these biases include a built-in conservative bias, inaccuracies in estimating the size of the affected universe, the effect of learning on

Table 3-4. (Continued)

POTENTIAL SOURCE OF ERROR	DIRECTION OF POTENTIAL BIAS FOR NET BENEFITS	LIKELY SIGNIFICANCE RELATIVE TO KEY UNCERTAINTIES ON NET BENEFITS ESTIMATE[1]
		reducing costs, the effect of innovation on reducing costs, and cost-reducing features of regulatory design. Some of these factors are discussed elsewhere in this table. The magnitude of these biases varies substantially, but in no case would we expect the overall impact to exceed five percent of overall net benefits.

[1] The classification of each potential source of error reflects the best judgment of the section 812 Project Team. The Project Team assigns a classification of "potentially major" if a plausible alternative assumption or approach could influence the overall monetary benefit estimate by approximately five percent or more; if an alternative assumption or approach is likely to change the total benefit estimate by less than five percent, the Project Team assigns a classification of "probably minor."

[2] U.S. Environmental Protection Agency Science Advisory Board, EPA-SAB-COUNCIL-ADV-07-002, "Benefits and Costs of Clean Air Act – Direct Costs and Uncertainty Analysis", Advisory Letter, June 8, 2007. Available at http://www.epa.gov/sab/pdf/council-07-002.pdf.

[3] For an analysis of the learning rates estimated in the empirical literature, see John M. Dutton and Annie Thomas, "Treating Progress Functions as a Managerial Opportunity," *Academy of Management Review*, Vol 9, No. 2, 1984.

[4] U.S. Department of Energy, Energy Information Administration, *Annual Energy Outlook 2005*, February 2005.

[5] The Project Team uses this maximum unit cost value in two ways. First, the Project Team assumes that local areas would not implement identified controls costing more than $15,000 per ton. Second, the Project Team assumes a cost of $15,000 per ton for unidentified controls.

[6] U.S. Environmental Protection Agency. *Regulatory Impact Analysis for the Particulate Matter NAAQS*. October, 2006.

[7] For detailed estimates of the costs of $PM_{2.5}$ NAAQS compliance, see E.H. Pechan and Associates, Inc. and Industrial Economics, Inc., *Direct Cost Estimates for the Clean Air Act Second Section 812 Prospective Analysis*, prepared for U.S. EPA, March 2009.

4. Air Quality Benefits

Air quality modeling links changes in emissions to changes in the atmospheric concentrations of pollutants that may affect human health and the environment. A crucial analytical step, air quality modeling is one of the more complex and resource-intensive components of the prospective analysis. This chapter outlines how we estimated future-year pollutant concentrations under both the *with-CAAA* and *without-CAAA* scenarios.

The first section of the chapter begins with a discussion of some of the challenges faced by air quality modelers and a brief description of the models we used in this analysis. The following section provides more details on the specific air quality modeling tools we deployed to estimate future-year ambient concentrations. This methodology section includes a description of how we use modeling results to adjust monitor concentration data and estimate ambient concentrations for years and scenarios where no monitoring yet exists – the projected and counterfactual (*without-CAAA*) target years and scenarios. The third section of this chapter summarizes the results of the air quality modeling and presents the expected effects of the CAAA on future-year pollutant concentrations. A brief discussion of the key uncertainties associated with air quality modeling concludes the chapter.

Overview of Approach

As we outlined in the First Prospective, air quality modelers face two key challenges in attempting to translate emission inventories into pollutant concentrations. First, they must model the dispersion and transport of pollutants through the atmosphere. Second, they must model pertinent atmospheric chemistry and other pollutant transformation processes. These challenges are particularly acute for those pollutants that are not emitted directly, but instead form through secondary processes. Ozone is the best example; it forms in the atmosphere through a series of complex, non-linear chemical interactions of precursor pollutants, particularly certain classes of volatile organic compounds (VOCs) and nitrogen oxides (NO_x). We faced similar challenges when estimating PM concentrations. Atmospheric transformation of gaseous sulfur dioxide and nitrogen oxides to particulate sulfates and nitrates, respectively, contributes significantly to ambient concentrations of fine particulate matter. In addition to recognizing the complex atmospheric chemistry relevant for some pollutants, air quality modelers also must deal with uncertainties associated with variable meteorology and the spatial and temporal distribution of emissions.

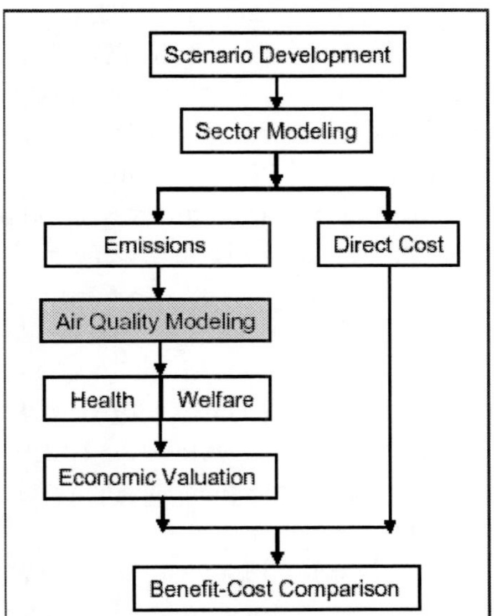

Air quality modelers and researchers have responded to the need for scientifically valid and reliable estimates of air quality changes by developing sophisticated atmospheric dispersion and transformation models. Some of these models have been employed in support of the development of federal clean air programs, national assessment studies, State Implementation Plans (SIPs), and individual air toxic source risk assessments. In this analysis, we focused our air quality modeling efforts on estimating the impact of with- and without-CAAA emissions on ambient concentrations of ozone, $PM_{10,}$ and $PM_{2.5}$, as well as acid deposition and visibility for each of the target years: 2000, 2010, and 2020. The focus on these pollutants is consistent with the result in the First Prospective that most of the quantified benefits of the CAAA are attributable to PM and ozone. The ideal model for this analysis is a single integrated air quality model capable of estimating ambient concentrations for all of these key pollutants throughout the U.S. In the prior First Prospective study, such a model had not yet been sufficiently developed and tested. This analysis is the first Section 812 prospective analysis to use an integrated modeling system, the Community Multiscale Air Quality (CMAQ) model, to simulate national and regional-scale pollutant concentrations and deposition. The CMAQ model (Byun and Ching, 1999) is a state-of-the-science, regional air quality modeling system that is designed to simulate the physical and chemical processes that govern the formation, transport, and deposition of gaseous and particulate species in the atmosphere.

The emissions data were processed for input to the CMAQ modeling using the Sparse-Matrix Operator Kernel Emissions (SMOKE) emissions processing system (CEP, 2004). The model-ready emission inventories for each scenario and year were then used to obtain base year and target year estimates of the key criteria pollutants, as well as many other species. The air quality modeling analysis was designed to make use of tools and databases that have recently been developed and evaluated by EPA for other national- and regional-scale air quality modeling studies. In particular, model-ready meteorological input files for 2002 were provided by EPA for use in this study. For fine particulate matter ($PM_{2.5}$) and related species, the CMAQ model was applied for an annual simulation period (January through December). A 36-km resolution modeling domain that encompasses the contiguous 48 states was used for the annual modeling. For ozone and related species, the CMAQ model was applied for a five-month simulation period that captures the key ozone-season months of May through September. Two 12-km resolution modeling domains (that when combined cover the key, ozone-significant areas of the contiguous 48 U.S. states) were used for the ozone-season modeling. Altogether, model-ready emission inventories were prepared and the CMAQ model was applied for a total of 21 simulations (comprising seven core scenarios and three modeling domains).

The outputs from the CMAQ model provide the basis for the calculation of health and ecological benefits of the CAA. The airborne criteria pollutants of interest include ozone and fine particulate matter ($PM_{2.5}$), where $PM_{2.5}$ consists of particles less than 2.5 microns in diameter. For health benefits analysis, it has become standard EPA practice to calibrate the CMAQ results monitor data, rather than use the CMAQ results directly – the process is sometimes called, "monitor and model relative adjustment." We follow that approach in this analysis as well, applying a tool called the Modeled Attainment Test Software (MATS) to develop and apply the calibration factors for particulate matter results relative to nearby monitors. For ozone, the MATS procedure is not necessary; instead we use an inverse distance squared weighting procedure called Enhanced Voronoi Neighbor Averaging (eVNA), which calibrates the CMAQ model ozone results by weighing data from monitors closer to the grid cell more heavily than monitors that are further away. The eVNA interpolation and model to monitor calibration process is accomplished within the BenMAP benefits analysis tool, which is described in Chapter 5. Visibility is also an air quality parameter of interest and this was calculated using a variety of the CMAQ output species. In addition, deposition of nitrogen and sulfur was also extracted from the model outputs. An overview of the modeling approach is provided in Figure 4-1, which summarizes the emissions processing and air quality components. The CMAQ modeling components and application of the MATS tool are explained in further detail in the next section.

Air Quality Modeling Tools Deployed

The CMAQ Modeling System

The Community Multiscale Air Quality (CMAQ) model is a state-of-the-science, regional air quality modeling system that can be used to simulate the physical and chemical processes that govern the formation, transport, and deposition of gaseous and particulate species in the atmosphere (Byun and Ching, 1999). The CMAQ tool was designed to improve the understanding of air quality issues (including the physical and chemical processes that influence air quality) and to support the development of effective emissions control strategies on both the regional and local scale. The CMAQ model was designed as a "one-atmosphere" model and this concept refers to the ability of the model to dynamically simulate ozone, particulate matter, and other species in a single simulation which captures interaction effects among these pollutants. In addition to addressing a variety of pollutants, CMAQ can be applied to a variety of regions with varying geographical, land-use and emissions characteristics, and for a range of different space and time scales.

The CMAQ model numerically simulates the physical processes that determine the magnitude, temporal variation and spatial distribution of the concentrations of ozone and particulate species in the atmosphere and the amount, timing, and distribution of their deposition to the earth's surface. The simulation processes include advection, dispersion (or turbulent mixing), chemical transformation, cloud processes, and wet and dry deposition. The CMAQ science algorithms are described in detail in Byun and Ching (1999).

The CMAQ model requires several different types of input files. Gridded, hourly emission inventories characterize the release of anthropogenic, biogenic and, in some cases, geogenic emissions from sources within the modeling domain. The emissions represent both low-level and elevated sources and a variety of source categories (including, for example, point, onroad mobile, nonroad mobile, area, and biogenic emissions). The amount, spatial distribution, and temporal distribution of each emitted pollutant or precursor species are key determinants to the resultant simulated air quality values.

The CMAQ model also requires hourly, gridded input fields of several meteorological parameters including wind, temperature, mixing ratio, pressure, solar radiation, fractional cloud cover, cloud depth, and precipitation. A full list of the meteorological input parameters is given in Byun and Ching (1999). The meteorological input fields are typically prepared using a data-assimilating prognostic meteorological model, the output of which is processed for input to the CMAQ model using the Meteorology-Chemistry Interface Processor (MCIP). The prescribed meteorological conditions influence the transport, vertical mixing, and resulting distribution of the simulated pollutant concentrations. Particular meteorological parameters, such as mixing ratio, can also influence the simulated chemical reaction rates.

Rainfall and near-surface meteorological characteristics govern the wet and dry deposition, respectively, of the simulated atmospheric constituents.

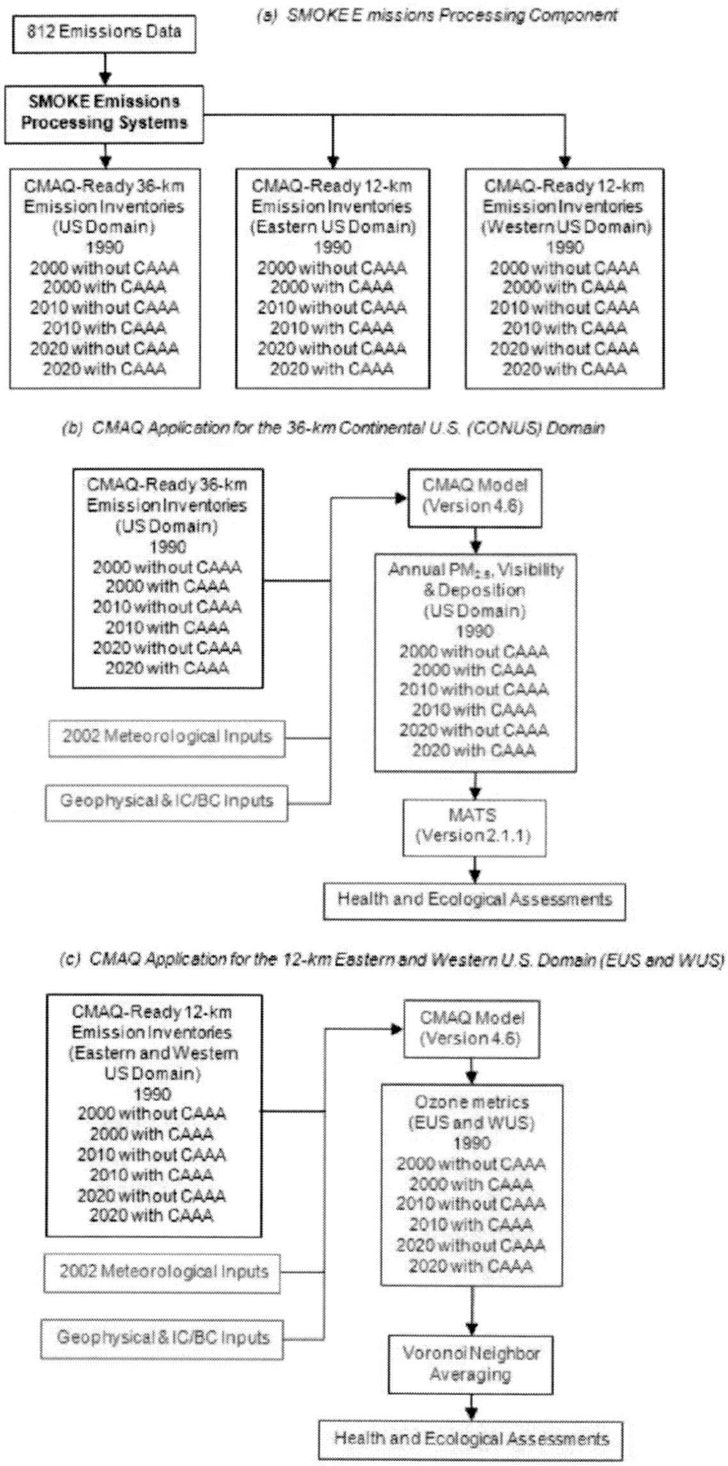

Figure 4-1. Schematic Diagram of Section 812 Air Quality Modeling Analysis.

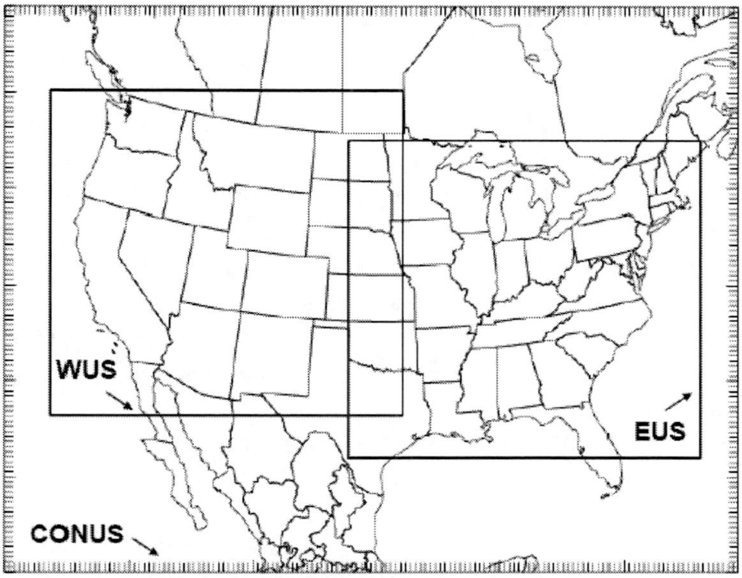

Note: Conus is the Continental us Grid Used for pm Modeling; Wus is the Western us Grid and eus is the Eastern US Grid Used for Ozone Modeling.

Figure 4-2. CMAQ Modeling Domains for the 812 Modeling Study.

Initial and boundary conditions (IC/BC) files provide information on pollutant concentrations throughout the domain for the first hour of the first day of the simulation, and along the lateral and top boundaries of the domain for each hour of the simulation. Photolysis rates and other chemistry related input files supply information needed by the gas-phase and particulate chemistry algorithms.[33]

CMAQ Application Procedures for the Second Prospective Analysis

This specific application of CMAQ includes modeling domain specification and key input files. The three modeling domains that were used for this analysis are shown in Figure 4-2.

The 36-km resolution continental U.S. (CONUS) domain is the large area that is covered by the outer grid box in Figure 4-2. The CONUS domain includes 148 x 112 grid cells (the total number of cells is 16,576). The tick marks denote the 36-km grid cells. For this domain, the model was run for the entire 2002 calendar year, using 2002 meteorology but varying the emissions inputs as outlined in each of the Second Prospective scenarios listed in Figure 4-1. In running the model, the annual simulation period was divided into two parts covering January through June and July through December, respectively. Each part of the simulation also includes an additional five start-up simulation days, which are intended to reduce the influence of uncertainties in the initial conditions on the simulation results.

The Eastern U.S. (EUS) domain is comprised of 213 x 188 grid cells (total = 40,044 cells) and the Western U.S. (WUS) domain includes 213 by 192 grid cells (total = 40,896 cells). Together these two domains cover most of the continental U.S. with 12-km horizontal resolution. There is some overlap in the central part of the country. For both the EUS and WUS domains, the CMAQ model was run for the months of May through September. This five-month period is intended to represent the ozone season – runs using this domain provide the ozone inputs for subsequent steps of the analysis. The seasonal simulation period was also divided into two parts covering May and June and July through September, respectively. Each part of the simulation also includes an additional ten start-up simulation days.

The 36- and 12-km resolution meteorological input files to support modeling in these domains were prepared using the Pennsylvania State University/National Center for Atmospheric Research (PSU/NCAR) Fifth Generation Mesoscale Model (MM5). The MM5 outputs were postprocessed by EPA for input to CMAQ using the Meteorology-Chemistry Interface Processor (MCIP) program. The meteorological input preparation methodology and some information on MM5 model performance are provided by Dolwick et al. (2007). Existing initial condition, boundary condition, land-use and photolysis rate input files prepared by EPA for use in CMAQ modeling for the selected modeling domains and simulation period were used.

After the initial CMAQ results were generated, the original primary PM emissions estimates generated for area and non-EGU point sources were found to be inaccurate due to two issues:

(1) As described in Chapter 2, some of the fine particulate emissions estimates derived from the 1990 NEI, on which the *without-CAAA* emissions estimates were based, were discovered to be inconsistent with those from the 2002 NEI, on which the *with-CAAA* emissions estimates were based.
(2) The original emissions estimates did not include application of transport factors for area source fine particulate emissions. These transport factors are county- specific adjustment factors that are applied to specific types of emissions estimates to account for the fact that only a fraction of total fugitive dust emissions remain airborne and are available for transport away from the vicinity of the source after localized removal (i.e., some of the particles are captured by the local vegetation or other surface obstructions).

To correct these two errors, we first made the necessary adjustments to the primary $PM_{2.5}$ emissions estimates for the affected non-EGU point and area sources, focusing on the $PM_{2.5}$ species that contribute most significantly to primary PM emissions: elemental carbon (EC), organic carbon (OC), and crustal material. We then calculated species- specific adjustment factors for the CMAQ data, re-compiled the species-specific estimates to generate an adjusted version of the original CMAQ results, and then generated new MATS input files. All details of the procedure are described in a memorandum prepared by the Project Team, which was reviewed in detail by the Council's Air Quality Modeling Subcommittee.[34]

Mats Procedure

Rather than using the direct CMAQ results as the basis for the health and ecological effects analyses, the Project Team conducted additional analyses using a speciated monitor and model calibration technique to generate $PM_{2.5}$ air quality estimates. The $PM_{2.5}$ estimates used in the Second Prospective health analysis were prepared using EPA's Modeled Attainment Test Software (MATS, Version 2.1.1, Build 807). MATS estimates quarterly mean $PM_{2.5}$ chemical component concentrations at monitor locations by conducting a Speciated Modeled Attainment Test (SMAT) analysis. MATS can also estimate quarterly mean concentration estimates for each $PM_{2.5}$ chemical component concentrations at all grid cells in a grid model such as CMAQ.

Five of the six MATS $PM_{2.5}$ concentration estimates for the Second Prospective scenarios were prepared using the MATS' spatial and temporal relative adjustment method. The MATS estimates for the 2000 *with-CAAA* scenario, which represents a historical year for which monitor data are available, used a spatial only relative adjustment method, relying on available monitor data and a single year of CMAQ modeling. The MATS procedure was not applied for the 1990 base year scenario.

MATS estimates the $PM_{2.5}$ concentrations in CMAQ grid cells by interpolating values from nearby monitors using the inverse distance squared weighting option in the Voronoi Neighbor Averaging (VNA) procedure in MATS. This is an algorithm that identifies a set of monitors close to the grid cell (called "neighbors") and then estimates the PM species concentration in that grid cell by calculating an inverse-distance weighted average of the monitor values (i.e., the concentration values at monitors closer to the grid cell are weighted more heavily than monitors that are further away). As noted above, for calibrating ozone model results to nearby monitors, only the VNA component of the procedure is used, because there is no need for the speciated interpolation approach required for PM.

The spatial MATS analysis conducted for the $PM_{2.5}$ estimates used the following input information:

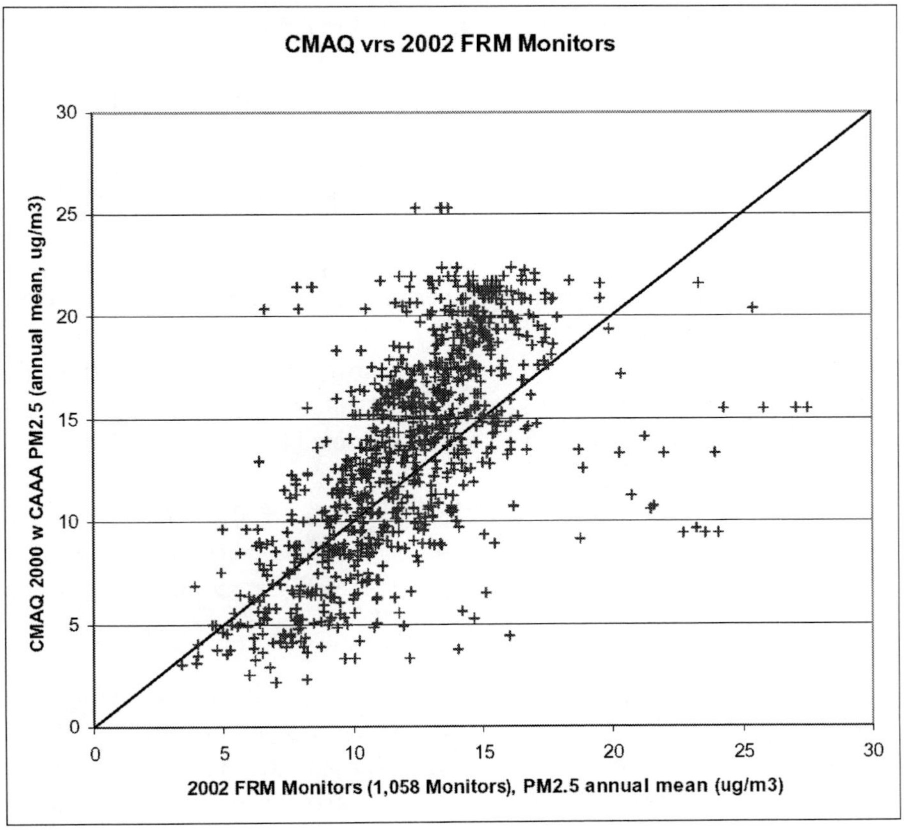

Figure 4-3. Scatter Plot of Direct CMAQ Estimates and 2002 $PM_{2.5}$ Federal Reference Method (FRM) Monitors.

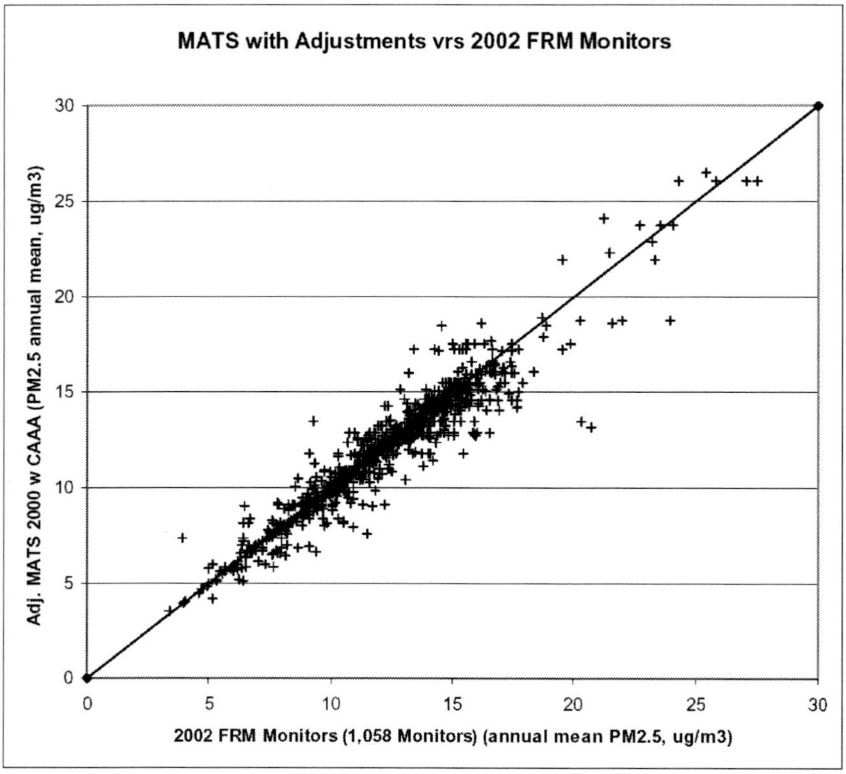

Figure 4-4 Scatter Plot of Mats-Adjusted CMAQ Estimates and 2002 $PM_{2.5}$ Federal Reference Method (FRM) Monitors.

- observed quarterly $PM_{2.5}$ data from 1,232 Federal Reference Method (FRM) monitors with sufficient data in 2002 – sufficient data is defined as at least one quarter of $PM_{2.5}$ data. The year 2002 was used because it corresponds to the vintage of the emissions estimates, which, as described in Chapter 2, were derived from the 2002 National Emissions Inventory;
- observed daily chemically speciated fine particle mass data from both the $PM_{2.5}$ Speciation Trends Network (STN) and the Interagency Monitoring of Protected Visual Environments (IMPROVE) network, providing a total of 273 monitors with sufficient data in 2002[35];
- speciated CMAQ estimates for 6 $PM_{2.5}$ species (SO4, NO3, elemental carbon, organic carbon, NH4, and crustal material) at the 36 kilometer $PM_{2.5}$ CMAQ grid cell level for each of the Second Prospective scenarios (from CMAQ speciated output data files).

The MATS procedure enables the use of monitor data to effectively calibrate the results of air quality modeling for use in subsequent steps of the analysis. To illustrate the effects of the MATS procedure, compare Figure 4-3, which is a scatter plot comparing the direct CMAQ results for those 1,058 $PM_{2.5}$ monitors with at least two quarters of data for 2002, and Figure 4-4, which is a similar scatter plot, comparing the MATS results to the same set of $PM_{2.5}$ monitors. The agreement between monitor and model values in Figure 4-4 is greatly improved by the MATS procedure.

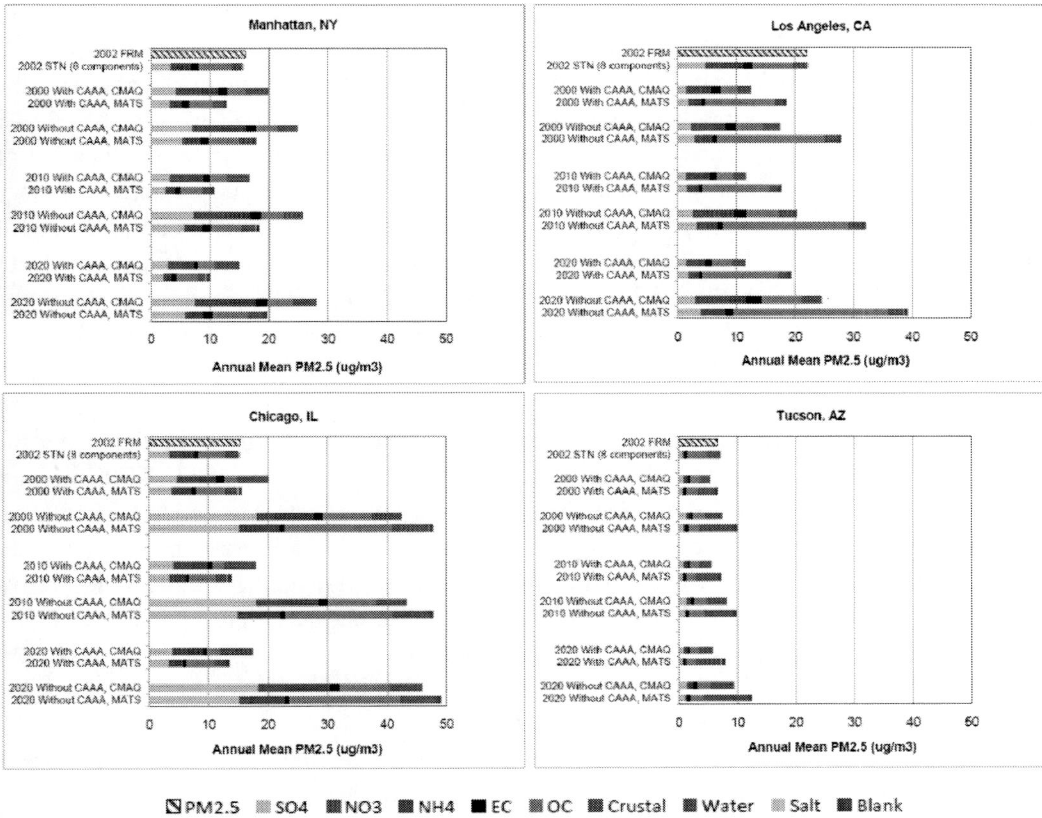

Figure 4-5. Comparison of CMAQ, Mats, and Monitor Data for Four Selected Cities.

Figure 4-5 provides a further illustration of the effect of the MATS procedure, and the importance of individual PM species in achieving an effective calibration of the CMAQ results to monitor data. The figure provides detailed species-specific CMAQ and MATS results for a CMAQ grid cell in the three largest cities and metropolitan areas in the US – New York, Los Angeles, and Chicago – and for Tucson, Arizona, a much smaller city but one for which one component of PM, crustal (shown in brown), plays a critical role in our analysis. For each city, the two leftmost bars provide the 2002 FRM and STN annual average $PM_{2.5}$ monitor data for a monitor of that type within the grid cell. FRM monitors provide only total $PM_{2.5}$ mass, while the STN monitors provide data for the seven PM species (plus estimated water) indicated at the bottom of each graph.[36] The remaining 12 bars in each panel show the CMAQ and MATS-adjusted results for the grid cell for the *with-CAAA* and *without-CAAA* scenarios, for target years 2000, 2010, and 2020.

The Manhattan panel in the upper left corner shows that both the FRM and STN monitors indicate a total PM concentration just greater than 15 $\mu g/m^3$. The next bar shows that the CMAQ data for the 2000 *with-CAAA* simulation overestimates the PM concentration, by about 4 $\mu g/m^3$. Comparing the 2002 STN bar with the 2000 *with-CAAA* CMAQ bar, we see that the CMAQ simulation overestimates most constituents in this location, compared to the monitor data, but underestimates organic matter (or OC, shown in green). The MATS procedure, applied to the STN and CMAQ data, generates species-specific scaling factors that result in a MATS-adjusted concentration for the 2000 *with-CAAA* scenario, shown in the next bar. As a result, the species-specific constituents in the MATS adjusted bar are in very nearly the same proportion as they appear for the STN monitor.

It would also appear from this figure that MATS "overcorrects" in Manhattan because the 2000 *with-CAAA* MATS bar is lower than the 2002 STN monitor bar. However, the MATS procedure is estimating the concentration at the center of the grid cell, not at the location of the STN monitor. In a 36 km grid cell, the monitor location can be many kilometers away from the center of the grid cell. MATS considers not only monitors in the same grid cell, but also the data at other nearby FRM and STN monitors, and makes a spatial interpolation to estimate concentrations at the grid centroid. The Manhattan STN monitor is near the intersection of four grid cells, which contain a total of 25 FRM and STN monitors, all of which influence the MATS result.

The remaining MATS estimates for Manhattan, for the 2000 *without-CAAA* and the 2010 and 2020 projections, are based on scaling of the corresponding CMAQ simulation by the species-specific factors developed from the 2000 *with-CAAA* to 2002 STN monitor comparison. The effect of MATS in Manhattan is to adjust the CMAQ simulation concentrations downward. Interestingly, the opposite is generally true in Los Angeles, because in that city CMAQ tends to underestimate the monitor data for 2002. The mix of species in both cities is similar in 2002, but strikingly different over time, particularly in the *without-CAAA* scenario, where organic carbon (shown in green) in Los Angeles derives from mobile sources, and sulfates (shown in yellow) in Manhattan derives from long-range transport from coal-burning electric generating units.

In Chicago, the effect of MATS is more complex, and the importance of considering PM species is highlighted. In the *with-CAAA* scenarios, MATS yields a downward adjustment to the CMAQ simulations, because the 2000 *with-CAAA* CMAQ simulation is higher than the 2002 STN monitor value. In the *without-CAAA* scenarios, however, there are much higher *emissions* of organic carbon, because certain OC emissions controls are not in place in the *without-CAAA* simulations that are in place in the *with-CAAA* scenario. Because CMAQ underestimates the ambient OC component in the 2000 *with- CAAA* (shown in green), the factor for OC that is applied to other scenarios yields an increase in concentration in the MATS-adjusted values. That increase is large enough to dominate the overall adjustment across all eight species, yielding an overall $PM_{2.5}$ mass increase for the *without-CAAA* scenarios relative to the CMAQ data.

The data for Tucson also illustrates the importance of the species-specific scaling factors. If it were not for changes to one PM species, crustal (shown in brown), there would be only a relatively modest difference between the *with-CAAA* and *without-CAAA* scenarios in future years. In Tucson the crustal component derives largely from construction activity, which in this relatively fast growing area of Arizona, and absent more stringent dust control measures, could become a larger issue in the projection years. CAAA controls on fugitive dust emissions in the construction sector, however, yield a substantial difference in this component of PM concentrations, when comparing the *with-CAAA* and *without-CAAA* scenario results. Other species differ much less across scenarios. In many other places like Tucson, the species-specific MATS procedure likely yields a more accurate projection of the impact of the CAAA than a calibration procedure that did not take into account the impact of these species-specific control strategies.

Air Quality Results

Particulate Matter

As mentioned above, the CMAQ modeling results for the 36-km continental U.S. (CONUS) modeling domain provide the basis for particulate matter air quality used in the calculation of PM-related health effects and to calculate visibility, as well as sulfur and nitrogen deposition. Summary results are presented in the maps in Figure 4-6 below, representing annual average concentrations across the CONUS domain for each of the seven scenario/target year combinations modeled. The rows of Figure 4-6 show modeled $PM_{2.5}$ concentrations for 2000, 2010, and 2020, contrasting the *without-CAAA* results on the left and the *with-CAAA* results on the right.

As the figure indicates, over the thirty-year 1990-2020 simulation period air quality is projected to worsen somewhat in the absence of CAAA regulations, particularly in the Midwest and California, but with CAAA regulations in place air quality is estimated to improve markedly as early as the year 2000 and to show continued improvements through 2020. In general, the *with-CAAA* results reflect a calibration of the 2002 model year results to monitor values, but as the accompanying Box 4-1 illustrates, such direct comparisons are not possible for the counterfactual *without-CAAA* results. We conclude for the analyses described in the text box that the *without-CAAA* results, with a few exceptions, seem to imply a return of air quality conditions comparable to those that prevailed in the 1980-1990 period prior to implementation of the CAAA. Such comparisons are limited, however, by the sparse $PM_{2.5}$ monitoring data for this period and the uncertainty in adjusting available monitor data for other species. Although the improvements attributed to the CAAA are nationwide, the most substantial gains are made in those areas that had the worst PM air quality in 1990, suggesting the CAAA has been and will continue to be effective in targeting improvements to the areas that would have experienced the worst air quality in the absence of the amendments.

BOX 4-1. EVALUATING THE *WITHOUT-CAAA* SCENARIO RESULTS

The two scenarios used in this study, the *with-CAAA* and *without-CAAA* scenarios, are designed to simulate and forecast air quality conditions in the US as we expect them to unfold with full implementation of the CAAA (the *with-CAAA*), and alternatively as if regulations authorized by the CAAA had not been implemented. In effect, the methods we use tie the *with-CAAA* scenario to monitored air quality in the year 2000, providing some measure of credibility for the air quality conditions reflected in our *with-CAAA* simulation. It is more difficult to evaluate the credibility of the *without-CAAA* scenario, because that scenario simulates hypothetical air quality conditions that cannot be observed. The plausibility of the *without-CAAA* scenario and its differences from the *with-CAAA* scenario nevertheless can be assessed through comparison to other similar air quality conditions.

One possible analog for conditions in the *without-CAAA* scenario is areas outside the US that have not implemented air quality regulations that match the stringency of those in the US. The problem with comparing US to non-US areas is the difficulty of standardizing factors which define air quality, such as meteorology, terrain, and the distribution of air pollutant emission sources. Another major challenge is that monitoring networks for fine particle species are sparse or not available for the annual average measure.

A preferable, though still imperfect, comparison is between *without-CAAA* forecasts and historical concentrations in US cities. A key issue arising for within-US comparisons is that prior to 1990 particulate matter monitors measured total suspended particulates (TSP), or PM_{10}, rather than $PM_{2.5.}$ The new PM standard is based on $PM_{2.5,}$ which is now recognized as better correlated with adverse health effects. $PM_{2.5}$ is therefore the focus of our air quality simulations. Furthermore, the ratios of TSP and/or PM_{10} to $PM_{2.5}$ vary considerably by location and over time, so a simple transformation of the available monitor data may not be reliable. Nonetheless, it is possible to find times and locations in the historical monitor data where at least two and sometimes all three of these measures were simultaneously collected, providing a means to estimate a time and location-specific ratio that can be used to infer $PM_{2.5}$ values. We use this type of information to develop estimates of historical $PM_{2.5}$ concentration in selected U.S. cities for comparison to our *without-CAAA* scenario projected values.

The table suggests that our estimates of *without-CAAA* $PM_{2.5}$ concentrations in New York, Pittsburgh, and Los Angeles are reasonably consistent with estimated historical concentrations in the 1980 to 1990 pre-CAAA period. In Chicago, however, the *without-CAAA* case yields estimates that are much higher than historical estimates. One reason may be the potentially strong influences of projected uncontrolled sulfur dioxide emissions from electric power plants near Chicago in the *without-CAAA* case. In the absence of Title IV these plants are projected in our study to use relatively high sulfur, locally mined coal and would not have been required to install scrubber technology.

(ANNUAL AVG MICROGRAMS PER CUBIC METER)	ESTIMATED $PM_{2.5}$ CONCENTRATIONS FOR THIS STUDY						ESTIMATED HISTORICAL $PM_{2.5}$		
	2000		2010		2020		1980 (EST)	1990 (EST)	MAXIMUM 1980-90
CITIES	W-CAAA	W/O-CAAA	W-CAAA	W/O-CAAA	W-CAAA	W/O-CAAA			
New York - Manhattan	12.9	20.6	10.9	21.0	10.0	22.1	N/A	22.4	N/A
New York - Queens/Brooklyn	13.2	24.8	11.0	25.2	10.1	26.7	N/A	21.5	N/A
Pittsburgh	14.0	19.2	11.0	19.7	10.0	20.3	29.3	22.3	29.8
Chicago	15.5	47.7	13.7	47.6	13.4	48.9	25.7	20.4	25.7
Los Angeles	18.5	25.5	17.1	29.7	17.5	35.5	38.5	29.4	41.9

Figure 4-7 makes the gains in 2020 more clear, by illustrating the differences in $PM_{2.5}$ concentrations between the *with-CAAA* and *without-CAAA* scenarios in 2020. The gains in some areas, particularly in the eastern half of the US, in California, and in urban centers nationwide, are dramatic, with reductions of more than 20 $\mu g/m^3$ in some areas. These are consistent with the large decreases in PM precursor emissions for those areas, described in

Chapter 2. In some of these areas, the *without-CAAA* scenario concentrations also reach high levels because of the absence of *without-CAAA* controls (see accompanying text box for a discussion of the *without-CAAA* scenario). There are also some surprisingly large reductions in a few less populous areas, such as, west central Idaho and central Virginia. The reductions in Idaho, as well as in a few other isolated areas of the rural West, are associated with CAAA requirements to limit emissions from agricultural burning operations. The reductions in central Virginia are attributable to local controls on a large coal-burning industrial boiler.

Some areas also experience modest increases in PM concentrations with the CAAA – these areas show up in light orange on the map. Some of the smallest estimated increases, less than 1 $\mu g/m^3$, can be introduced by the MATS adjustment procedure, particularly when the locations are far from monitors and/or have very low modeled or monitored concentrations of a PM species. We interpret very small increases such as these as effectively "no change" so adjusted the map legend to group these cells with others where are small benefits.[37] There remain five cells with disbenefits greater than 1 $\mu g/m^3$. The three cells of these five with the smallest disbenefit estimates did not have disbenefits in the CMAQ modeling – we therefore conclude that the disbenefit result was introduced by the MATS procedure.

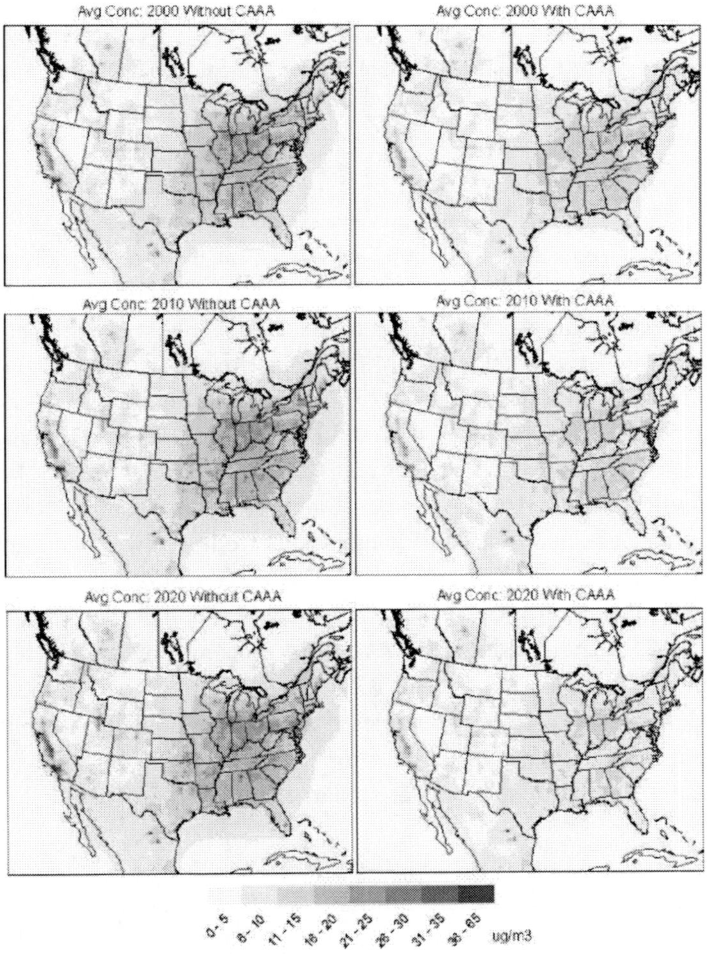

Figure 4-6. CMAQ Simulated and Mats Adjusted Annual Average $PM_{2.5}$ Species Concentration (Micrograms Per Cubic Meters) for the Conus Domain Outputs for the 1990 to 2020 Period.

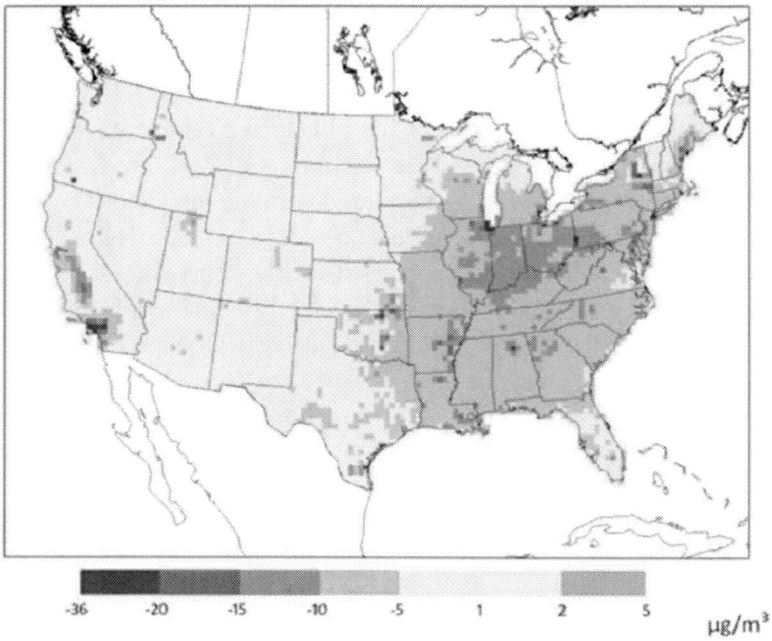

Figure 4-7. Difference in CMAQ Simulated Mats Adjusted Annual Average $PM_{2.5}$ Species Concentration (Micrograms per Cubic Meter) for the Conus Domain: 2020 with-CAAA Minus 2020 without-CAAA Scenarios.

In the remaining two cells, we conclude that implementation of the CAAA led to negative benefits, associated with actual increases in emissions resulting in the with-CAAA case relative to the *without-CAAA* case. The largest disbenefit, of 4.1 $\mu g/m^3$, is in the northwestern corner of New Mexico, in the cell which includes the Four Corners Power Plant, one of the largest coal-burning power plants in the West. The emissions data indicate sulfur dioxide emissions for that plant that are 14,000 tons greater in the 2020 *with-CAAA* case, probably as a combined result of changes in dispatch and sulfur content of coal for this plant, which as of December 2010 does not have a sulfur scrubber. The other cell shows a disbenefit of 1.25 $\mu g/m^3$, and is located in Sweetwater County in south central Wyoming, which includes the Pacificorp-Jim Bridger Power Plant. The air quality result here is also attributable to a difference in sulfur dioxide emissions from a power plant, in this case 2,000 tons greater in the 2020 *with-CAAA* scenario. The dispatch of this unit appears to be identical in both scenarios, so the result is most likely attributable to a marginal reallocation of higher sulfur coal. Note that, as indicated in the *with-CAAA* maps in Figure 4-6, these are areas that nonetheless would continue to experience $PM_{2.5}$ concentrations below the 15 $\mu g/m^3$ $PM_{2.5}$ annual standard. These relatively modest and geographically limited exceptions notwithstanding, it is clear that by 2020 the air quality benefits of the CAAA in reducing ambient concentrations of particulate matter are large and widespread.

Ozone

Figures 4-8 through 4-11 present similar CMAQ output data for ozone, with two important differences: (1) the ozone results are reported for the Eastern (EUS) and Western (WUS) 12-km modeling domains; and (2) the results presented are the average of daily maximum 8-hour ozone concentration, in ppb, over the course of a modeled ozone season

(May 1 through September 30). The average daily 8-hour maximum may seem like an odd metric for evaluating ozone concentrations, but because this is the metric used in epidemiological estimation of mortality risks of ozone this metric is closely correlated with the major mortality incidence and economic benefits associated with ozone precursor controls. Results for the Eastern US are in Figures 4-8 and 4-9, and for the Western US in Figures 4-10 and 4-11.

For the Eastern US, Figure 4-8 shows a similar pattern for ozone as was illustrated for particulate matter in Figure 4-6. That is, while there are relatively modest increases in ozone concentrations in the absence of the CAAA, the *with-CAAA* maps on the right side of the graphic show significant and widespread gains in air quality throughout the region, with air quality benefits increasing over time. By 2020, Figure 4-9 shows that the difference in ozone concentrations is large in most areas of the east, with gains as large as 30 ppb for this simulated day.

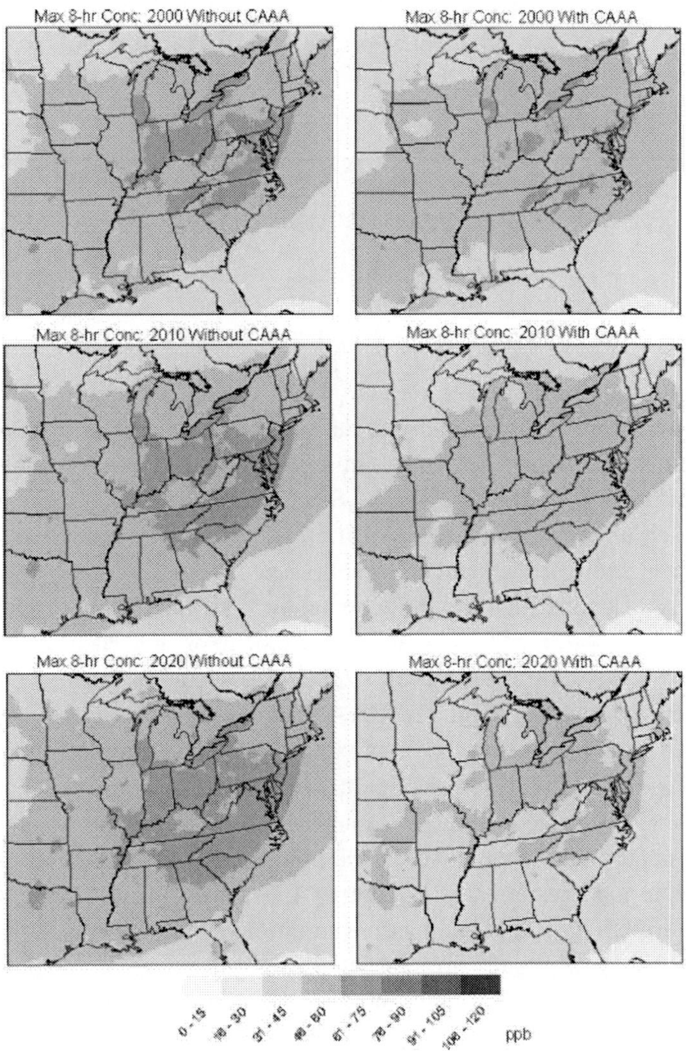

Figure 4-8. Cmaq Simulated and VNA Adjusted Daily Maximum 8-Hour Ozone (PPB) for the EUS Domain.

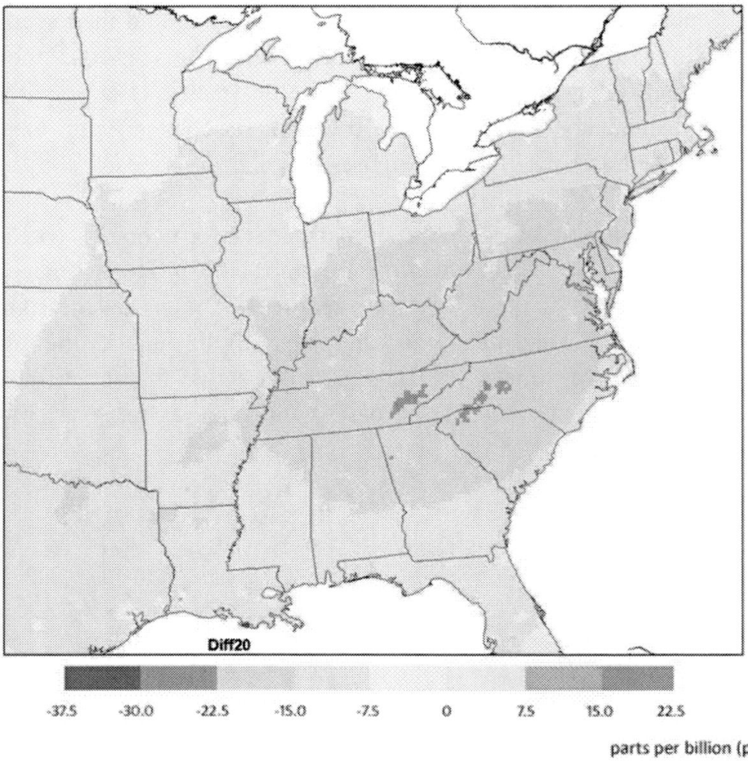

Figure 4-9. Difference in Simulated Daily Maximum 8-Hour Ozone Concentration (PPB) for the EUS Domain for 15 July: 2020 with-CAAA Minus 2020 without- CAAA Scenarios.

Two other patterns in Figure 4-9 are also worth noting. First, although the region-wide benefits of the CAAA are large, in many urban areas concentrations in the *with-CAAA* case are higher than in the *without-CAAA* case, in some cases near the Gulf Coast and in New York City by as much as 15 to 20 ppb. Second, some of the areas with the largest improvements, such as those in the heart of the Midwest, include pockets of much smaller gains, particularly in some urban centers. In both cases, these results are not unexpected. The complex chemistry of ozone includes a phenomenon known as "NO_x-scavenging", whereby nitrogen oxides, while participating as an ozone precursor, can also serve to scavenge or reduce ozone, particularly during the peak ozone season and in urban centers where ozone levels might otherwise be quite high. The CAAA, in reducing the nitrogen oxide precursors, may in some cases reduce ozone on a regional level while leading to much smaller reductions or even increases in ozone in the center of certain urban areas. This effect explains both these results. Nonetheless, as Figure 4-9 makes clear, the overall area (and population exposed) of ozone reductions is far greater than the corresponding areas with ozone increases.

Ozone results in the Western US, in Figures 4-10 and 4-11, indicate a similar pattern to those for the Eastern US when examining concentrations in urban areas, although in the West the largest ozone air quality gains are restricted to a smaller area, centered in the areas in California that have historically struggled with ozone attainment. In addition, in the Western US there are some more extensive areas in Figure 4-11 with ozone disbenefits attributed to the CAAA, particularly in Los Angeles.[38] Another interesting result, not shown in Figure 4-10, is that we estimate that ozone concentrations will actually increase from 1990 to 2000 in most parts of California, in both the *withoutCAAA* and *with-CAAA* scenarios, before reductions in 2010 and 2020 bring ambient levels below those seen in 1990, at least in most areas. This result is largely attributable to the longer attainment deadlines for the severe non-attainment areas in California – our scenario assumes that emissions will increase for some period before aggressive regional mobile source tailpipe standards and non-road fuel and engine standards, and local-scale ozone attainment plans, have their full effect later in our simulation period.

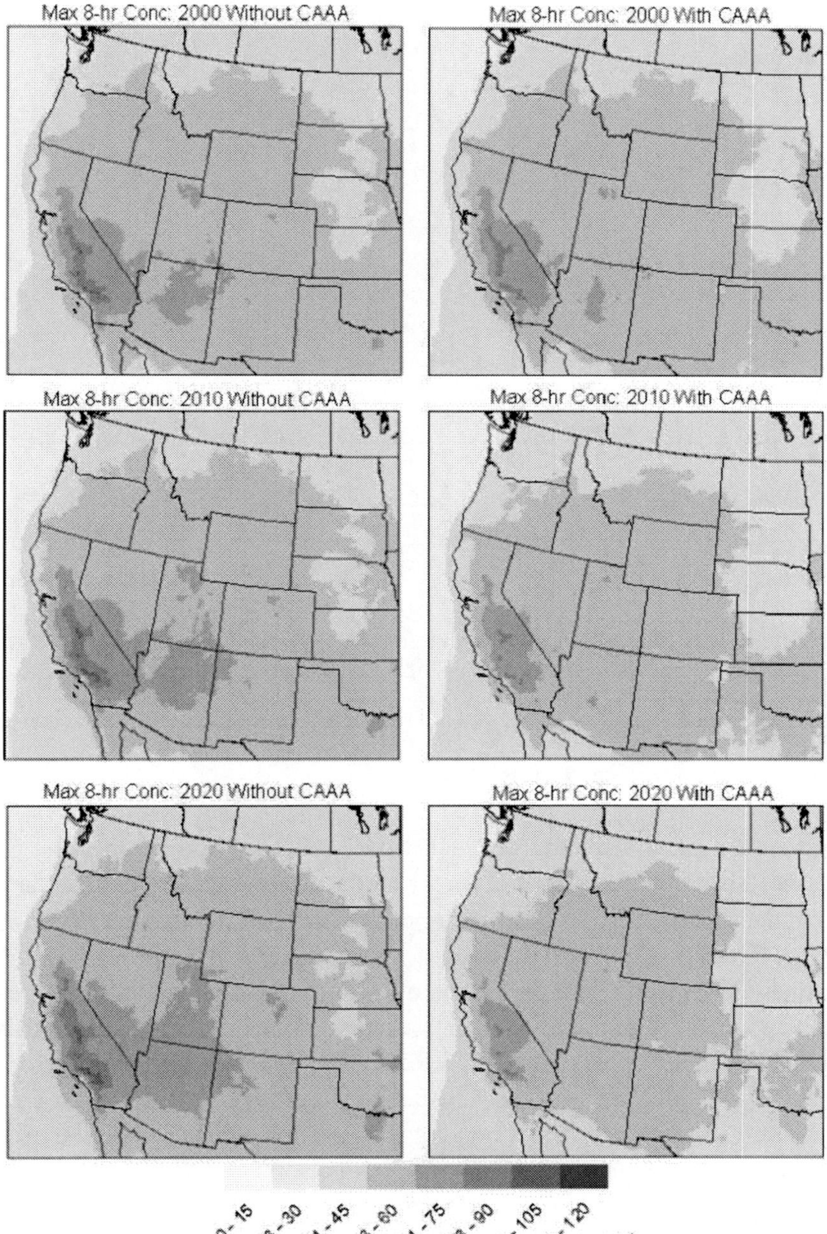

Figure 4-10. CMAQ Simulated and VNA Adjusted Daily Maximum 8-Hour Ozone (PPB) for the WUS Domain.

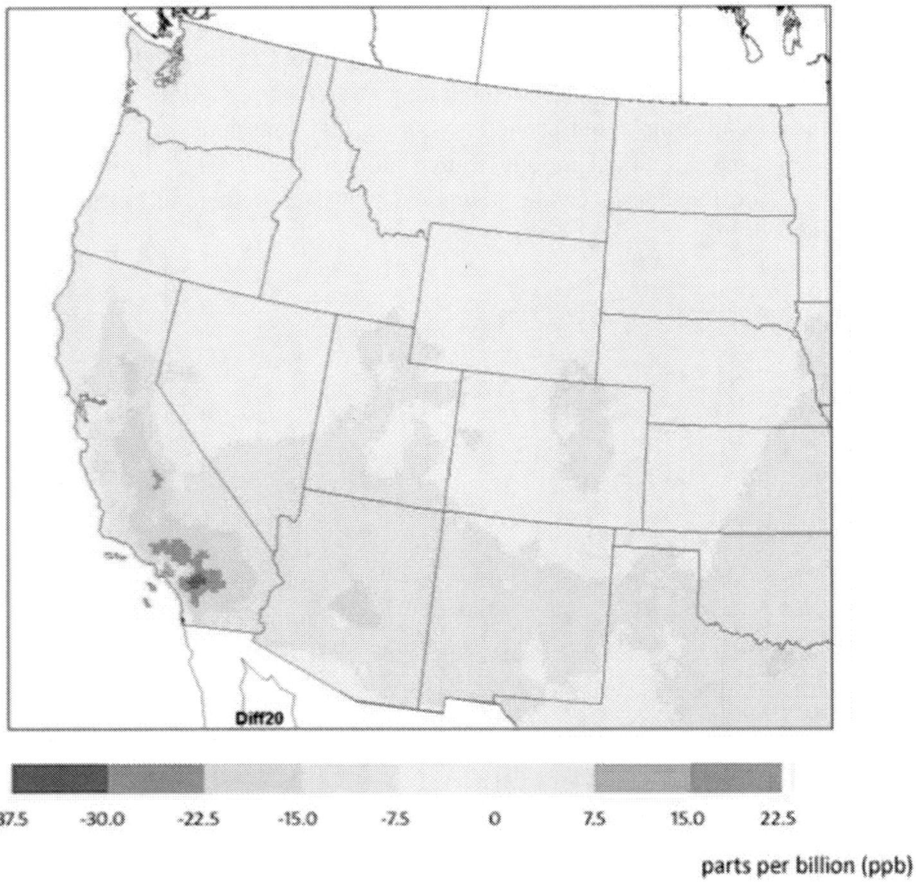

Figure 4-11. Difference in Simulated Daily Maximum 8-Hour Ozone Concentration (PPB) for the WUS Domain for 15 August: 2020 with-CAAA Minus 2020 without- CAAA Scenarios.

Uncertainty in Air Quality Estimates

Unlike the air quality modeling conducted over a decade ago for the first Section 812 prospective analysis, which used two different models for ozone and particulate matter, the modeling conducted for the Second Prospective analysis utilized EPA's Community Multiscale Air Quality (CMAQ) model, a "one-atmosphere" model that simulates the chemical formation, transport, and deposition of ozone and particulate matter together in one comprehensive system.[39] The use of this comprehensive air quality modeling system provides a consistent platform for evaluating the expected responses to changes in precursor emissions, reducing many of the uncertainties which pertained in the First Prospective as a result of the limited ability of the models to capture important interaction effects among the ozone and PM precursor pollutants.

Nonetheless, air quality modeling is a complex process and, as such, involves many uncertainties. We provide a summary of some of the more important classes of air quality modeling uncertainties in Table 4-1 below. These include a known meteorological bias in the 1 2-km eastern MM5 domain, which leads to a general tendency to underestimate the monthly observed precipitation; uncertainties in secondary organic aerosol (SOA) chemistry which lead to underestimation of SOA formation in the CMAQ simulations; issues in the detailed CMAQ modeling of some PM precursors; reliance for ozone modeling on a 12-km grid, suggesting NO_x inhibition of ambient ozone levels may be under-represented in some urban areas; and some emissions estimation geographic scale/resolution issues. In all cases but the ozone grid resolution and modeling of SOA formation, the effect of these uncertainties on our estimate of net benefits is of uncertain direction. In addition, in all but one case, modeling of SOA formation, we believe the impact of these uncertainties is probably minor, or of an influence less than five percent of the total net benefits, based on current information. Use of the CMAQ model platform, which has been evaluated in many contexts and used extensively by EPA for broad regulatory analyses such as the Second Prospective, has been a major factor enhancing our understanding of the impact of air quality modeling exercises such as this.

Another factor contributing to our understanding of key uncertainties is that the air quality modeling analysis conducted for the second Section 812 prospective study used national-scale modeling databases originally prepared by EPA for use in other recent modeling exercises conducted to support national rulemaking, including the latest available meteorological and other input databases (for 2002). Given that the modeling databases were originally prepared and utilized by EPA in other analyses, a comprehensive performance evaluation was not undertaken as part of this Section 812 prospective analysis; though the overall projections were assessed using the Atmospheric Model Evaluation Tool (AMET), which showed bias and error statistics for our results were within the acceptable range for model performance.[40] As noted in Table 4-1, biases or uncertainties could be manifest in the simulated concentration fields due to the use of the 36- and 12-km resolution grids, which might not be sufficiently detailed to resolve certain sub-grid scale processes in portions of the modeling domain. All air quality modeling exercises are affected by inherent uncertainties in model formulation, meteorological inputs, and emission inventory estimates. Nevertheless, the modeling was conducted following current EPA guidelines and in a manner consistent with EPA approaches/practice for similar national-scale modeling exercises.

One factor identified in Table 4-1 involves uncertainties associated with corrections to the air quality outputs completed coincident with the Council review of the study outputs. These corrections, reflecting the need to adjust some categories of direct fine particulate emissions for the *without-CAAA* scenario, and to incorporate adjustments to take account of processes that remove fugitive dust from the ambient air at or close to the source of emissions, owing to the effect of forests, vegetation, and urban structures on fugitive dust, were necessary because of issues identified through quality control assessments the Project Team completed. As noted in the table, we believe these factors have been addressed through carefully designed *post-hoc* adjustment of the CMAQ results, however in both cases it would have been preferable to have made the adjustments prior to running the CMAQ model. Resource and time limitations unfortunately prevented the Project Team from re-estimating the CMAQ results to account for these adjustments.

Perhaps surprisingly, our assessment is that only one of these factors, uncertainty in secondary organic aerosol formation, constitutes a major source of uncertainty. This result could reflect our inability to apply alternative quantitative air quality modeling tools in this already resource-intensive step in the analytic chain, although it is also clear that the CMAQ model best reflects the state-of-the-art for the type of national scale air quality modeling necessary to support this benefit-cost analysis. As we discuss in Chapter 7, the overall contribution of this step in the analytic chain to uncertainty in net benefits, compared to other steps, may be considerably less, because of the ability to calibrate model results to monitor values for at least the year 2000 *with-CAAA* scenario. It is worth noting, however, that as a whole the air quality modeling process very likely contributes a greater than 10 percent uncertainty, of indeterminate direction, to the overall uncertainty in benefits estimates. In addition, it is clear there are uncertainties introduced by the *ex post* adjustment of some primary PM emissions estimates and the procedure used to re-calibrate the CMAQ air quality to account for this emissions adjustment. Although we argue that the overall effect of this source of uncertainty on the net benefits is probably minor, in some locations ambient PM from primary PM emissions can be more important than secondarily formed fine particles. Overall, we believe that our application of the MATS monitor calibration procedure, which provides a speciated calibration to ensure better agreement between air quality modeling results and comparable monitor data, provides the best attainable consistency between our air quality simulation results and monitored values – the ability to calibrate our results to detailed monitor data in this step of the analytic chain provides considerably greater confidence that our results are "ground-truthed" as much as possible to real world conditions.

Table 4-1. Key Uncertainties Associated with Air Quality Modeling.

POTENTIAL SOURCE OF ERROR	DIRECTION OF POTENTIAL BIAS FOR NET BENEFITS	LIKELY SIGNIFICANCE RELATIVE TO KEY UNCERTAINTIES ON NET BENEFITS ESTIMATE*
Unknown meteorological biases in the 12-km western and 36-km MM5 domains due to the lack of model performance evaluations.	Unable to determine based on current information.	Probably minor. Other evaluations using 2002 and similar meteorology and CMAQ have shown reasonable model performance, but significant effects on nitrate results in western areas with wintertime $PM_{2.5}$ problems.
Known metrological biases in the 12-km eastern MM5 domain. MM5 has a cold bias during the winter and early spring, and has a general tendency to under-estimate the monthly observed precipitation. MM5's under prediction was greatest in the fall and least in the spring months.	Unable to determine based on current information.	Probably minor. These biases would likely influence $PM_{2.5}$ formation processes, which was modeled on the 36-km domain.
Secondary organic aerosol (SOA) chemistry. CMAQ version 4.6 has known biases (underprediction) in SOA formation.	Underestimate.	Possibly major. The modeling system underpredicts SOA, which has both biogenic and anthropogenic components. Reductions in NOx can reduce both biogenic and anthropogenic SOA and reductions in VOC will reduce anthropogenic SOA.

Table 4-1. (Continued)

POTENTIAL SOURCE OF ERROR	DIRECTION OF POTENTIAL BIAS FOR NET BENEFITS	LIKELY SIGNIFICANCE RELATIVE TO KEY UNCERTAINTIES ON NET BENEFITS ESTIMATE*
		Since both of these precursors are significantly impacted by the CAAA, there may be large benefits from SOA related reductions that are not currently captured by the modeling system.
The CMAQ modeling relies on a modal approach to modeling $PM_{2.5}$ instead of a sectional approach. The modal approach is effective in modeling sulfate aerosol formation but less effective in modeling nitrate aerosol formation than the sectional approach.	Unable to determine based on current information.	Probably minor in the eastern U.S. where annual $PM_{2.5}$ is dominated by sulfate. Potentially major in some western U.S. areas where $PM_{2.5}$ is dominated by secondary nitrate formation.
Limited model performance evaluation of CMAQ for 2002.	Unable to determine based on current information.	Probably minor. While a comprehensive model evaluation was not completed, the overall results of the CMAQ runs for the Second Prospective were assessed using AMET, and bias and error statistics were within acceptable ranges. Further, our application of the MATS procedure provides further assurance that air quality results used in the subsequent health assessments are consistent with available monitor data.
Ozone modeling relies on a 12-km grid, suggesting NO_x inhibition of ambient ozone levels may be underrepresented in some urban areas. Grid resolution may affect both model performance and response to emissions changes.	Unable to determine based on current information.	Probably minor. Though potentially major ozone results in those cities with known NO_x inhibition, ozone benefits contribute only minimally to net benefit projections in this study. Grid size affects chemistry, transport, and diffusion processes, which in turn determine the response to changes in emissions, and may also affect the relative benefits of low-elevation versus high-stack controls.
Emissions estimated at the county level (e.g., low-level source and motor vehicle NO_x	Unable to determine based on current information.	Probably minor. Potentially major for estimation of ozone, which depends largely on VOC

Table 4-1. (Continued)

POTENTIAL SOURCE OF ERROR	DIRECTION OF POTENTIAL BIAS FOR NET BENEFITS	LIKELY SIGNIFICANCE RELATIVE TO KEY UNCERTAINTIES ON NET BENEFITS ESTIMATE*
and VOC emissions) are spatially and temporally allocated based on land use, population, and other surrogate indicators of emissions activity. Uncertainty and error are introduced to the extent that area source emissions are not perfectly spatially or temporally correlated with these indicators.		and NO_x emissions; however, ozone benefits contribute only minimally to net benefit projections in this study.
Use of MATS relative response factors to calculate changes in $PM_{2.5}$	Indeterminate	Probably minor. Using MATS, air quality modeling results were projected in a "relative" sense. In this approach, the ratio of future year model predictions to base year model predictions are used to adjust ambient measured data up or down depending on the relative (percent) change in model predictions for each location. The use of ambient data as part of the calculation helps to reduce uncertainties in the future year predictions, especially if the absolute model concentrations are over-predicted or under-predicted.
Modeling artifacts created by changes in emissions inventory estimation methods between the 1990 inventories used for the *without-CAAA* scenario and the 2002 inventories used for the *with-CAAA* scenarios were mitigated through application of adjustment factors for primary PM from non-EGU point sources, and for the certain subsectors of area sources, in the *without- CAAA* case. Application of these adjustments may result in overestimated or underestimated changes in primary PM contributions to ambient concentrations for these particular sources.	Unable to determine based on current information.	Probably minor. While primary PM can make a significant contribution to ambient $PM_{2.5}$ in some locations, secondarily formed fine particles dominate the estimates for ambient concentration change in this analysis. In addition, the effect of the inventory adjustments was to significantly reduce the differentials between the control and counterfactual scenarios, implying any residual error is more likely to reflect an underestimation bias than an overestimation bias, particularly since the non-EGU primary PM reductions were adjusted to a scenario differential of zero.

Table 4-1. (Continued)

POTENTIAL SOURCE OF ERROR	DIRECTION OF POTENTIAL BIAS FOR NET BENEFITS	LIKELY SIGNIFICANCE RELATIVE TO KEY UNCERTAINTIES ON NET BENEFITS ESTIMATE*
Adjustments to take account of processes that remove fugitive dust from the ambient air at or close to the source of emissions, owing to the effect of forests, vegetation, and urban structures on fugitive dust. Analysis of the chemical species collected by ambient air samplers suggests that the modeling process may overestimate PM-2.5 from fugitive dust sources by as much as an order of magnitude, if not adjusted for this effect. The Project Team incorporated adjustments post-CMAQ modeling but prior to use of PM air quality estimates in subsequent steps of the analysis.	Unable to determine based on current information.	Probably minor. If adjustment factors had been applied as part of the CMAQ modeling, evidence suggests the entrainment effect would have been adequately accounted for. The largely linear processes of direct PM emissions to air quality suggest that our *post-hoc* adjustment should also be adequate to account for this factor. Further assurance that this factor has been accounted for is our application of the MATS monitor calibration procedure, which provides a speciated calibration to ensure better agreement between air quality modeling results and comparable monitor data, and the fact that the adjustment applies to both scenarios, further mitigating the impact of this source of uncertainty.

* The classification of each potential source of error is based on those used in the First Prospective Analysis. The classification of "potentially major" is used if a plausible alternative assumption or approach could influence the overall monetary benefit estimate by approximately 5% or more; if an alternative assumption or approach is likely to change the total benefit estimate by less than 5%, the classification of "probably minor" is used.

5. Estimation of Human Health Effects and Economic Benefits

A large portion of the overall benefits of the Clean Air Act Amendments (CAAA) of 1990 are due to human health benefits from improved air quality. As part of the Second Prospective analysis of these amendments, we identified and, where possible, estimated the magnitude of health benefits Americans are likely to realize in future years as a result of the CAAA. We express these health benefits as avoided cases of air pollution-related health effects, such as premature mortality, heart disease and respiratory illness. Human health benefits of the 1990 CAAA can be attributed to reduced emissions of criteria pollutants (Titles I through IV), and reduced emission of ozone depleting substances (Title VI), however as highlighted in Chapter 1 the Second Prospective focuses primarily on human health effects attributed to the reduction of criteria pollutants, and within that category, health benefits associated with reduced exposure to fine particulate matter ($PM_{2.5}$) and ozone, as these are the largest contributors to the overall health benefits estimates.

The goal in a benefit-cost analysis such as the Second Prospective is to develop estimates of the monetary value of benefits wherever possible – doing so facilitates comparison and aggregation of monetized health benefits across endpoints. Therefore, we assigned a dollar value to avoided incidences of each health effect. We obtained valuation estimates from the economic literature and report them in "dollars per case avoided." We report each of the monetary values of benefits applied in this analysis in terms of a central estimate and a probability distribution around that value. The statistical form of the probability distribution varies by endpoint.

This chapter presents an overview of our approach to modeling changes in adverse health effects and applying monetary value to these benefits, summarizes the results for major health effect categories and discusses key uncertainties related to the analysis. As noted above, the chapter focuses primarily on the human health effects associated with exposure to criteria pollutants, however we also present the methodology and results of a case study of health benefits from a single air toxic pollutant (benzene) for a particular area of the United States (the Houston metropolitan area).

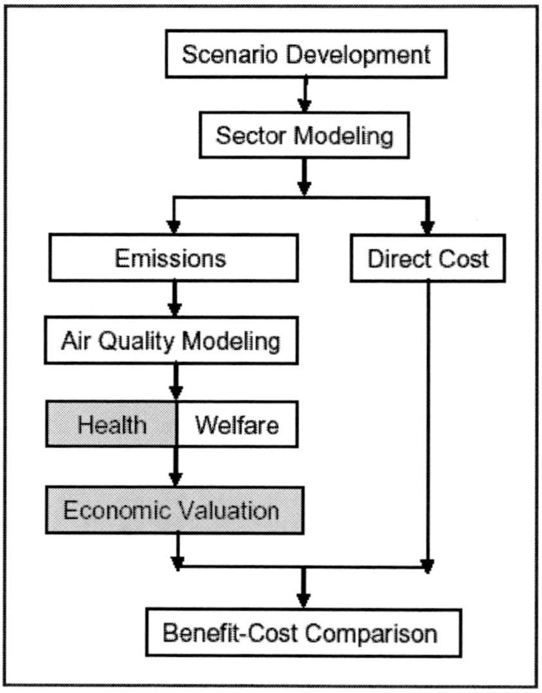

Overview of Approach

We estimate the impact of the CAAA on human health by analyzing the difference in the expected incidence of adverse health effects between a "*with-*" and a "*without-CAAA*" regulatory scenario. As described in Chapter 1, the *without-CAAA* scenario assumes no further controls on criteria pollutant emissions aside from those already in place in 1990, while the *with-CAAA* scenario assumes full implementation of the 1990 CAAA. The analysis uses a sequence of linked analytical models to estimate health benefits, also described in Chapter 1, which includes forecasts of implementation activities undertaken in response to the CAAA, estimates of pollutant emissions associated with each scenario (see Chapter 2) and air quality modeling of criteria pollutant emissions under each scenario (see Chapter 4).

Estimating health effects benefits from air quality modeling results involves three key steps, described in greater detail below. The first step involves estimating the exposure of individuals to air pollutants. Although exposure to air pollutants can occur in both outdoor and indoor environments, for our purposes it is appropriate to focus on outdoor air pollution concentrations as a measure of human exposure. The main reason is that, in the second step of our approach, estimating the human response to exposure, the exposure measures used in the epidemiological studies used to derive human response are typically based on outdoor concentrations. These "concentration-response functions" were developed to relate outdoor concentrations to changes in the incidence of health effects and mortality in response to pollutant exposure. The third step, valuation of avoided human health risk, is accomplished by application of estimates from the literature to characterize unit values per case avoided.

A critical tool in EPA's analyses of health benefits is the Environmental Benefits Mapping and Analysis Program (BenMAP), developed and continuously maintained by EPA's Office of Air and Radiation.[41] BenMAP is capable of accepting a wide range of air quality inputs, and then performing exposure analysis that includes calibration of model results to monitor data for historical years, assessing the changes in health effects incidence resulting from those exposures, and estimating the monetized value of those avoided health effects. Health effects in BenMAP are based on differences in two scenarios of exposure, and health effects and valuation estimates reflect the implications of the difference in exposure across scenarios, rather than absolute estimates of incidence associated with in any given scenario. BenMAP required three types of inputs for this analysis: 1) forecasted changes in air quality from the *without-CAAA* to the *with- CAAA* scenarios in 2000, 2010 and 2020; 2) health impact functions that quantify the relationship between the forecasted changes in exposure and expected changes in adverse health effects; and 3) health valuation functions that assign a monetary value to changes in specific health effects. We describe each of these inputs in greater detail below. The outputs of BenMAP for this analysis include central estimates and distributions of health effects incidence and valuation, at the national and county level, for each of the three target years of analysis.

The Project Team also estimates two other outputs related to avoided premature mortality attributed to the CAAA: life-years lost, and changes in life expectancy. EPA developed a separate model, the Population Simulation model, to generate these outputs. As described below, the population simulation approach provides some advantages over the BenMAP model in terms of simulation of the dynamic effects of mortality across a population through time, but also has several significant disadvantages relative to BenMAP in terms of the spatial resolution of pollutant exposure estimates. As a result, the population simulation approach operates as a supplement to the BenMAP-based primary estimates for selected measures of the impact of reducing risks of premature mortality.

Exposure Assessment

As described in Chapter 4, the Project Team used the Community Multi-scale Air Quality (CMAQ) integrated modeling system to simulate the physical and chemical processes that govern the formation, transport, and deposition of gaseous and particulate species in the atmosphere. The CMAQ results serve as the basis of the air quality inputs required for BenMAP. For particulate matter, the CMAQ model was applied for an annual simulation period (January through December) and utilized a 36-km resolution modeling domain that encompasses the contiguous 48 states. For ozone and related species, the CMAQ model was applied for a five-month simulation period that captures the key ozone-season months of May through September, and used two 12-km resolution modeling domains (that when combined cover the contiguous 48 U.S. states). We also described in Chapter 4 the adjustment of the CMAQ results generated by combining those results with observed monitoring data, using a method known as the monitor and model relative adjustment procedure. This technique was applied for the PM estimates using a program called the Modeled Attainment Test Software (MATS) (see Chapter 4 for a detailed description of this process). The resulting 36 km grid cell concentrations for PM were then used as inputs for BenMAP. For ozone, a similar adjustment process was completed, but the analysis was done directly within BenMAP, using the enhanced Voronoi Neighbor Averaging (eVNA) procedure.[42] The eVNA and MATS procedures provide gridded estimates of outdoor air quality at the same grid resolution as the CMAQ results. These procedures also provide a means for calibrating model results in those grid cells where no monitors exist, combining both model results with nearby monitor results to yield a "surface" of air quality that avoids the problems with direct extrapolation of results from monitors not located within a grid cell boundary.

Health Impact Functions

Health impact functions estimate the change in a health endpoint of interest, such as hospital admissions, for a given change in ambient pollutant concentration. A standard health impact function has four components: 1) the size of the potentially affected population; 2) a baseline incidence rate for the health effect (obtained from a source of public health statistics, such as the Centers for Disease Control, or sometimes from an epidemiological study itself); 3) a concentration-response (C-R) function (derived from epidemiological studies), which relates the change in the number of individuals in a population exhibiting a "response" to a change in pollutant concentration experience to the size of the exposed population; and 4) the estimated change in the relevant pollutant concentration. The first three of these components are discussed in further detail below. The fourth is generated through the air quality modeling and exposure estimation procedure discussed above.

Potentially Affected Populations

Health benefits resulting from the CAAA are related to the change in air pollutant exposure experienced by individuals. Because the expected changes in pollutant concentrations vary from location to location, individuals in different parts of the country may not experience the same level of health benefits. This analysis apportions benefits among individuals by matching the change in air pollutant concentration in a grid cell with the size of the population that experiences that change.

BenMAP incorporates 2000 U.S. Census Bureau block-group population data to determine the specific populations potentially affected by ozone and $PM_{2.5.}$ For future years (2010 and 2020), BenMAP scales the 2000 Census-based population estimates using the ratio of forecasted and 2000 county-level population estimates provided by Woods and Poole (2007).[43]

Baseline Incidence Rates

Baseline incidence rates are needed to convert the relative changes of a health effect in relation to a specific change in air pollution, which are reported in epidemiological studies, into the number of avoided cases. For instance, an epidemiological study might report that for a 10 ppb decrease in daily ozone levels, hospital admissions decrease by three percent. This estimate must then be multiplied by a baseline incidence rate (i.e., an estimate of the number of cases of the health effect per year) and the total population to determine how this three percent decrease translates into the number of fewer cases.

For this analysis, we used nationally-representative age-specific incidence and prevalence rates, where available, for each health endpoint. We obtained these data from a variety of sources, such as the CDC, the National Center for Health Statistics and the American Lung Association. Information from individual epidemiological studies was used if data from other sources were not available, as these data are often specific to the study population and location and therefore may not be as nationally representative.[44] For future years, mortality rates are projected based on available Bureau of the Census projections – other projected baseline incidence rates are generated to be consistent with the projections of population growth incorporated into BenMAP.

Concentration-Response Functions

We calculate the benefits attributable to the CAAA as the avoided incidence of adverse health effects. Such benefits can be measured using C-R functions specific to each health effect. C-R functions are equations that relate the change in the number of individuals in a population exhibiting a "response" (in this case an adverse health effect such as respiratory disease) to a change in pollutant concentration experienced by that population.

$PM_{2.5}$ and ozone have been associated with a number of adverse health effects in the epidemiological literature, such as premature mortality, hospital admissions, emergency room visits, and respiratory and cardiovascular disease. The published scientific literature contains information that supports the estimate of some, but not all, of these effects. Thus, it is not possible currently to estimate all of the human health benefits attributable to the CAAA. In addition, for some of the health effects we do quantify, the current economic literature does not support the estimation of the economic value of these effects. Table 5-1 lists the human health effects of these pollutants that have been identified, indicating which have been included in our benefits estimates and those that we did not quantify. See Chapter 2 of *Health and Welfare Benefits Analyses to Support the Second Section 812 Benefit-Cost Analysis of the Clean Air Act*, for a specific list of the C-R functions used for each health endpoint.

We rely on the most recently available, published scientific literature to ascertain the relationship between air pollution and adverse human health effects. We use a set of criteria outlined in Table 5-2 to evaluate potential studies to use as the basis for the C-R function. These criteria include consideration of whether the study was peer-reviewed, the study design and location, and characteristics of the study population, among others. In addition, we consider the input of the Council advising EPA for this study, as well the specific advice of the Health Effects Subcommittee (HES) of the Council, which explicitly focused on the health effects estimation component of the study. Overall, the selection of C-R functions for benefits analysis is guided by the goal of achieving a balance between comprehensiveness and scientific defensibility.

Table 5-1. Human Health Effects of Ozone and $PM_{2.5}$

Pollutant/Effect	Quantified and Monetized in Base Estimates[a]	Unquantified Effects[g,h]—Changesin:
PM/Health[b]	Premature mortality based on both cohort study estimates and on expert elicitation[c,d] Bronchitis: chronic and acute Hospital admissions: respiratory and cardiovascular Emergency room visits for asthma Nonfatal heart attacks (myocardial infarction) Lower respiratory symptoms Minor restricted-activity days Work loss days Asthma exacerbations (asthmatic population) Upper Respiratory symptoms (asthmatic population) Infant mortality	Subchronic bronchitis cases Low birth weight Pulmonary function Chronic respiratory diseases other than chronic bronchitis Morphological changes Altered host defense mechanisms Cancer Non-asthma respiratory emergency room Visits UVb exposure (+/-)[e]
Ozone/Health[f]	Premature mortality: short-term exposures Hospital admissions: respiratory Emergency room visits for asthma Minor restricted-activity days School loss days Outdoor worker productivity	Cardiovascular emergency room visits Asthma attacks Respiratory symptoms Chronic respiratory damage Increased responsiveness to stimuli Inflammation in the lung Premature aging of the lungs Acute inflammation and respiratory cell damage Increased susceptibility to respiratory infection Non-asthma respiratory emergency room Visits UVb exposure (+/-)[e]

a Primary quantified and monetized effects are those included when determining the primary estimate of total monetized benefits of the alternative standards.

b In addition to primary economic endpoints, there are a number of biological responses that have been associated with PM health effects including morphological changes and altered host defense mechanisms. The public health impact of these biological responses may be partly represented by our quantified endpoints.

c Cohort estimates are designed to examine the effects of long-term exposures to ambient pollution, but relative risk estimates may also incorporate some effects due to shorter term exposures (see Kunzli et al., 2001 for a discussion of this issue).

d While some of the effects of short-term exposure are likely to be captured by the cohort estimates, there may be additional premature mortality from short-term PM exposure not captured in the cohort estimates included in the primary analysis.

e May result in benefits or disbenefits.

f In addition to primary economic endpoints, there are a number of biological responses that have been associated with ozone health including increased airway responsiveness to stimuli, inflammation in the lung, acute inflammation and respiratory cell damage, and increased susceptibility to respiratory infection. The public health impact of these biological responses may be partly represented by our quantified endpoints.

g The categorization of unquantified health effects is not exhaustive.

h Health endpoints in the unquantified benefits column include both a) those for which there is not consensus on causality and b) those for which causality has been established but empirical data are not available to allow calculation of benefits.

Table 5-2. Summary of Considerations Used in Selecting C-R Functions.

CONSIDERATION	COMMENTS
Peer-Reviewed Research	Peer-reviewed research is preferred to research that has not undergone the peer-review process.
Study Type	Among studies that consider chronic exposure (e.g., over a year or longer), prospective cohort studies are preferred over ecological studies because they control for important individual-level confound-ding variables that cannot be controlled for in ecological studies.
Study Period	Studies examining a relatively longer period of time (and therefore having more data) are preferred, because they have greater statistical power to detect effects. More recent studies are also preferred because of possible changes in pollution mixes, medical care, and lifestyle over time. However, when there are only a few studies available, studies from all years will be included.
Population Attributes	The most technically appropriate measures of benefits would be based on impact functions that cover the entire sensitive population but allow for heterogeneity across age or other relevant demographic factors. In the absence of effect estimates specific to age, sex, preexist-ing condition status, or other relevant factors, it may be appropriate to select effect estimates that cover the broadest population to match with the desired outcome of the analysis, which is total national-level health impacts. When available, multi-city studies are preferred to single city studies because they provide a more generalizable representation of the C-R function.
Study Size	Studies examining a relatively large sample are preferred because they generally have more power to detect small magnitude effects. A large sample can be obtained in several ways, either through a large popu-lation or through repeated observations on a smaller population (e.g., through a symptom diary recorded for a panel of asthmatic children).
Study Location	U.S. studies are more desirable than non-U.S. studies because of potential differences in pollution characteristics, exposure patterns, medical care system, population behavior, and lifestyle.
Pollutants Included in Model	When modeling the effects of ozone and PM (or other pollutant combinations) jointly, it is important to use properly specified impact functions that include both pollutants. Using single-pollutant models in cases where both pollutants are expected to affect a health outcome can lead to double-counting when pollutants are correlated.
Measure of PM	For this analysis, impact functions based on $PM_{2.5}$ are preferred to PM_{10} because of the focus on reducing emissions of $PM_{2.5}$ precursors, and because air quality modeling was conducted for this size fraction of PM. Where $PM_{2.5}$ functions are not available, PM1 0 functions are used as surrogates, recognizing that there will be potential downward (upward) biases if the fine fraction of PM_{10} is more (less) toxic than the coarse fraction.
Economically Valuable Health Effects	Some health effects, such as forced expiratory volume and other tech-nical measurements of lung function, are difficult to value in monetary terms. These health effects are not quantified in this analysis.
Non-overlapping Endpoints	Although the benefits associated with each individual health endpoint may be analyzed separately, care must be exercised in selecting health endpoints to include in the overall benefits analysis because of the possibility of double- counting of benefits.

Epidemiological studies provide the basis for the C-R functions used in the health impact functions for assessing benefits of the CAAA. These studies also provide an indication of a portion of the uncertainty associated with the C-R function, by reporting a confidence interval around the mean value, which we use to derive a low, central and high estimate of avoided cases. However, this range only represents the statistical error in the estimates, which is related to the study population size and frequency of outcome. Several other sources of uncertainty exist in the relationship between ambient pollution and the health outcomes, including model uncertainty, potential confounding by factors that are both correlated with the health outcome and each other, and potential misclassification of the study population exposures. For a full list of uncertainties related to application of a C-R function to estimate benefits, see the Uncertainty section of this chapter and the Second Prospective Uncertainty Report, *Uncertainty Analyses to Support the Second Section 812 Benefit-Cost Analysis of the Clean Air Act*.

EPA recently conducted an expert elicitation (EE) study, which is the formal elicitation of subjective judgments, in order to more fully characterize the uncertainty surrounding the $PM_{2.5}$/mortality C-R function. This study allowed experts to consider and integrate several sources of uncertainty in the form of a probability distribution of the C-R function. As discussed further below, the EE study results helped to inform our selection of a primary C-R function to estimate avoided premature mortality due to CAAA-related $PM_{2.5}$ exposure reductions.

Avoided premature mortality is the largest contributor to the monetized health benefits of $PM_{2.5}$ and ozone. Therefore, we describe below in further detail the specific C-R functions selected to quantify CAAA-related avoided deaths.

PM Mortality C-R Function

The estimated relationship between particulate matter exposure and premature mortality is one of the most important parameters in the overall quantified and monetized benefit estimate for this study. An extensive base of literature exists to support development of the C-R function linking fine particulate matter exposure with premature mortality. Our knowledge of both the potential biological mechanisms linking $PM_{2.5}$ exposure with mortality and the potential magnitude of this effect has grown since the First Prospective was completed as the result of continued research and follow-up of existing study populations. Both short-term and long-term epidemiological studies have been conducted to examine the PM/mortality relationship. Short-term exposure studies attempt to relate short-term (often day-to-day) changes in PM concentrations and changes in daily mortality rates up to several days after a period of elevated PM concentrations. Longterm exposure studies examine the potential relationship between longer-term (e.g., annual) changes in exposure and annual mortality rates. Although positive, significant results have been reported using both of these study types, we rely exclusively on longterm studies to quantify PM mortality effects. This is because cohort studies are able to discern changes in mortality rates due to long-term exposure to elevated air pollution concentrations. This provides a better match to the benefits of air pollution control programs under the CAAA, which are also focused on reducing long-term exposure. These effect estimates may also include some of the mortality changes due to short-term peak exposures.[45] Therefore, the use of C-R functions from long-term studies is likely to yield a more complete assessment of the effect of PM on mortality risk.

Among long-term PM studies, we prefer those using a prospective cohort design to those using an ecologic or population-level design. Prospective cohort studies follow individuals forward in time for a specified period, periodically evaluating each individual's exposure and health status. Population-level ecological studies assess the relationship between population-wide health information (such as counts of daily mortality) and ambient levels of air pollution. Prospective cohort studies are preferred because they are better at controlling a source of uncertainty known as "confounding." Confounding is the mis-estimation of an association that results if a study does not control for factors that are correlated with both the outcome of interest (e.g., mortality) and the exposure of interest (e.g., PM exposure). For example, smoking is associated with mortality. If populations in high PM areas tend to smoke more than populations in low PM areas, and a PM exposure study does not include smoking as a factor in its model, then the mortality effects of smoking may be erroneously attributed to PM, leading to an overestimate of the risk from PM. Prospective cohort studies are better at controlling for confounding than ecologic studies because the former follow a group of individuals forward in time and can gather individual-specific information on important risk factors such as smoking.

Two major prospective cohort studies have been conducted in the U.S.: the American Cancer Society (ACS) study and the Six Cities study. These two cohorts are large, produce consistent results, provide broad geographic coverage and have been independently reexamined and reanalyzed. Strengths of the ACS study over the Six Cities study include greater geographic coverage (50 U.S. cities) and larger sample size. However, a key limitation of this study is a recruitment method that led to a study population with higher income, more education, and greater proportion of whites than the general U.S. population. In addition, available monitoring data was often assigned to all of the individuals within a large metropolitan area, potentially allowing for exposure misclassification.[46] Both of these limitations could imply that the ACS results are potentially biased low. The Six Cities study included a more representative sample of subjects within each community and set up monitors purposefully for the study. It was therefore able to assign exposures at a finer geographic scale. However, this study only included six cities and therefore may not be representative of the entire U.S. population, mix of air pollutants, and other potentially important factors.

The extensive epidemiological literature is complemented by EPA's 2006 expert elicitation (EE) study that asked 12 leading experts in PM health effects to integrate this pool of knowledge with the various sources of uncertainty that hinder our ability to precisely identify the true mortality impact of a unit change in annual $PM_{2.5}$ concentration (IEc, 2006). The results of the expert elicitation study showed three important findings: first, that advances in the scientific literature led many of the interviewed scientists to espouse greater confidence in the linkage between $PM_{2.5}$ exposure and mortality; second, that many of the experts believed that the central estimate of the mortality effect was considerably higher than the Pope et al. (2002) result used in the First Prospective; and third, that most of the experts' uncertainty distributions of the mortality effect reflected a much wider range of possible values, both high and low, than were used in the First Prospective study. The expert elicitation study does not, however, provide an integrated distribution across all 12 experts of possible values for the PM-mortality C-R function.

Based on consultations with the Council's Health Effects Subcommittee (HES), the 812 Project Team developed a distribution of C-R function coefficients (i.e., the percent change in annual all-cause mortality per one μg/m^3 change in annual average $PM_{2.5}$) for use in the PM-mortality C-R function for the Second Prospective study. This distribution is rooted in the epidemiological studies that most inform our understanding of the PM- mortality C-R function, but reflects the broader findings of the EE study. We based the primary C-R coefficient estimate of the Second Prospective study on a Weibull distribution with a mean of 1.06 percent decrease in annual all-cause mortality per one μg/m^3. This mean is roughly equidistant between the results of the two most well-studied PM cohorts, the ACS cohort (0.58, as derived from Pope et al., 2002) and the Six Cities cohort (1.5, as derived from Laden et al., 2006), both of whose results have been robust to continued follow-up and extensive re-analysis. Half of the coefficient values in this distribution fall between these two studies, one-quarter are higher than the Laden mean estimate, and one-quarter are lower than the Pope mean estimate; however all coefficient values are greater than zero. This distribution is consistent with the EE results described above, showing considerable support for higher values based on results from more recent studies (e.g., the Laden et al. (2006) Six Cities follow-up) and concerns cited by the Council HES that the ACS cohort results may underestimate the true effect. The use of all positive values is consistent with both the increased confidence in a causal link between $PM_{2.5}$ exposure and mortality shown in the EE study and the lack of evidence in general to support a threshold for mortality effects of $PM_{2.5}$ in the U.S. population.[47]

The results of two recently published cohort studies provide additional support for the selection of the Weibull distribution as the primary estimate for the PM Mortality C-R function. The first is a large retrospective cohort study of over 13 million Medicare participants (i.e., those aged 65 and above) throughout the US (Eftim et al. 2008; Zeger et al. 2008). When the entire Medicare cohort was analyzed, authors found a 6.8 percent change in annual all-cause mortality in the eastern US (95% CI: 4.9-8.7) and a 13.2 percent change in the central US (95% CI: 9.5-16.9) per 10 μg/m^3 change in the long-term (six-year) average annual $PM_{2.5}$. There was no association found in the western US (Zeger et al., 2008). These results are similar to the interquartile range of the Weibull distribution selected for the primary estimate for the Second Prospective. An analysis restricted to those living in the locations corresponding to the ACS and Six Cities cohort study analyses yielded percent changes in annual all-cause mortality per 10 μg/m^3 of $PM_{2.5}$ of 10.9 (95% CI: 9.0-12.8) and 20.8 (95%CI: 14.8-27.1) respectively, which are somewhat higher than the estimates reported in the original studies (Eftim et al., 2008).[48] One possible explanation for this difference is the lack of control for lifestyle factors in the analyses by Eftim et al., such as smoking, potentially leading to confounded results.

The second study is a prospective cohort of female nurses in the Northeastern and Midwestern regions of the US (Puett et al. 2008 and 2009). An increase of 10 μg/m^3 of $PM_{2.5}$ in the previous year was associated with a 26 percent increase in annual all-cause mortality (a hazard ratio of 1.26 with a 95% CI ranging from 1.02 to 1.54).[49] This estimate is at the upper end of our primary estimate Weibull distribution (roughly equivalent to the 95th percentile). However, this study covered only two regions of the country and included only females and therefore may not be generalizable to the general population of the US.

A final topic concerns EPA's choice to estimate avoided mortality and morbidity associated with reductions in fine particles using estimates of changes in exposure to fine particle mass as the exposure input in the damage function. The implication of this approach is that we assume that all fine particles, regardless of their chemical composition, are equally potent per unit concentration in producing premature mortality and other health outcomes. If it could be shown that fine particle species exhibit significantly differentiated toxicity, then from a benefits analysis perspective, treatment of all fine particle species as equally toxic would lead to biased benefits estimates, because the composition of fine particle mass varies over space and time, as do the fine particle reductions resulting from different air pollutant control strategies. We believe that these biases would likely be minor in an analysis such as the 812 study, which evaluates a blended particle reduction strategy targeting multiple particle types across the entire spectrum of control programs authorized under the Clean Air Act Amendments. Nonetheless, we conducted a careful evaluation of the potential for characterizing uncertainty in the differential toxicity of the components of fine particle pollution.

There exists a limited but growing literature addressing the health effects of various fine particle components, including sulfate, nitrate, elemental carbon (EC), organic carbon (OC), and metals.[50] A number of epidemiological studies, mostly time-series studies, have associated one or more of the components of fine particle pollution individually with mortality; however, so far no clear picture has emerged to implicate specific components as being consistently more toxic than fine particles in general or to classify any individual components of fine particle pollution as non-toxic. However, the epidemiological evidence base is limited by the high correlations among many fine particle components (and between those components and fine particles as a whole). It is difficult to corroborate this evidence toxicologically, given the fact that human exposure to single particle components is not a realistic scenario. The literature base continues to expand, but significant investments in both epidemiological and toxicological research are needed to understand the potentially complex systems of particle interactions that may be responsible for the observed health effects of fine particle pollution.

Thus, while treatment of all fine particle components as equally toxic may lead to biases in benefits estimates, we also acknowledge that any arbitrary assumption about the differential toxicities of specific fine particle types may also lead to biases in benefits estimates. Any of these biases may mask important spatial variation in the distribution of benefits of Clean Air Act programs across the U.S. due to regional variation in fine particle species mixes, which could affect selection of the most health beneficial measures to meet Clean Air Act requirements such as the National Ambient Air Quality Standards. However, the "equal toxicity" fine particle approach is rooted in both biological considerations (i.e., the importance of particle size to toxicity) and in largely consistent findings across an extensive set of epidemiological studies conducted across countries, states, and cities that show $PM_{2.5}$ concentrations are associated with increased mortality and morbidity rates. This consistency of results across a variety of fine particle mixes in different locations implies an equivalence of risk resulting from exposure to fine particle masses with different concentrations of component species. We conclude that the current evidentiary base from the epidemiological and toxicological literatures supports the use of an equal toxicity assumption for the present study, especially since the fine particle pollution reductions estimated herein reflect a variety of fine particle mixtures across different locations and time frames. Furthermore, we conclude that current information does not support specification of alternative concentration-response functions that would be both scientifically sound and useful for development of policy-relevant insights.

To provide further confidence that the results presented in this chapter are not likely to be substantially affected by the possibility that $PM_{2.5}$ species exhibit differential toxicity, the Project Team developed and evaluated estimates of the overall population-weighted exposure to PM species. The results are presented in Table 5-3 below, and graphically in the two panels of the accompanying Figure 5-1. The results in Figure 5-1 indicate that the population-weighted composition of fine particulate matter is affected by the control strategies applied in the CAAA, but the changes are relatively modest.[51] We therefore conclude that, even if species-specific toxicity estimates could be derived from the existing literature, applying them in this study would not have a large effect on the mortality results presented later in this chapter.

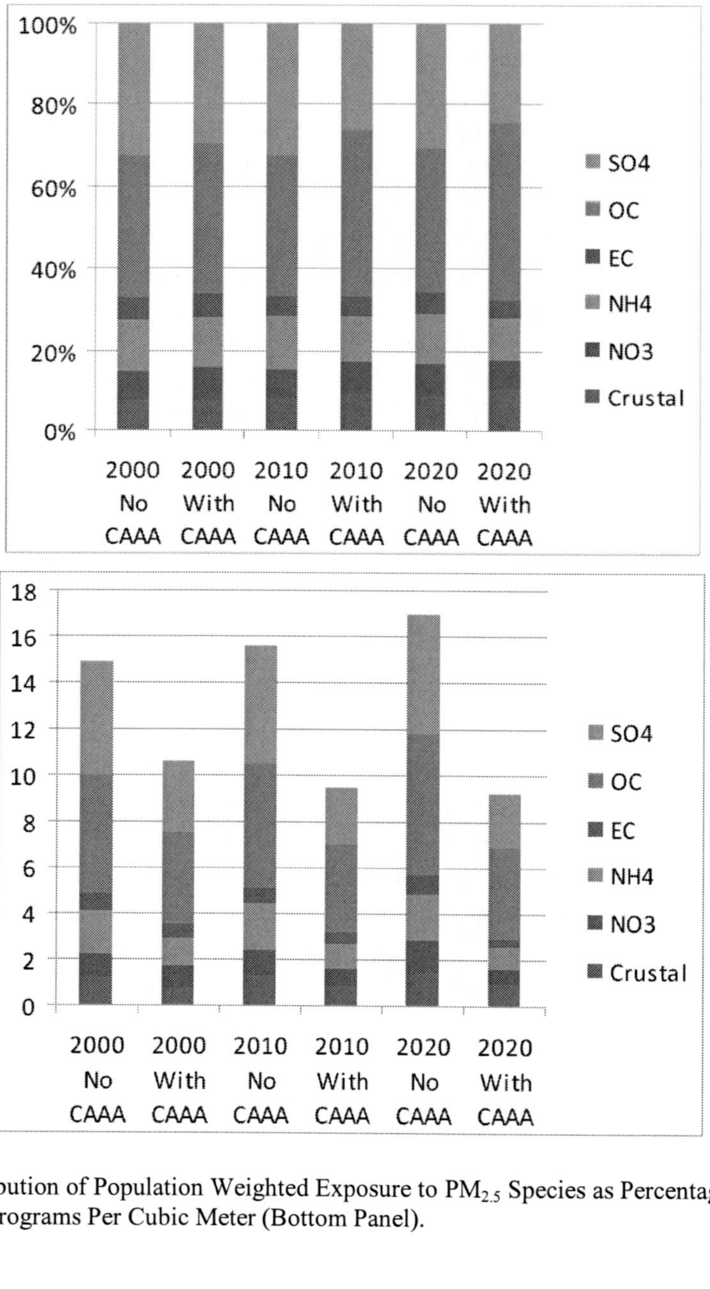

Figure 5-1. Distribution of Population Weighted Exposure to $PM_{2.5}$ Species as Percentage of Total (Top Panel) and in Micrograms Per Cubic Meter (Bottom Panel).

Table 5-3. Estimated Population Weighted Exposure For $PM_{2.5}$ Species (Micrograms Per Cubic Meter).

	2000 NO CAAA	2000 WITH CAAA	2010 NO CAAA	2010 WITH CAAA	2020 NO CAAA	2020 WITH CAAA
Crustal	1.18	0.82	1.27	0.86	1.51	0.96
NO_3	1.06	0.89	1.17	0.81	1.32	0.69
NH_4	1.87	1.26	1.96	1.03	2.05	0.92
EC	0.74	0.62	0.77	0.48	0.9	0.41
OC	5.18	3.94	5.36	3.86	6.02	3.99
SO_4	4.84	3.11	5.02	2.48	5.17	2.22

Ozone Mortality C-R Function

Several recent epidemiological studies suggest that ozone exposure likely contributes to premature mortality.52 Epidemiological data are also supported by recent human and animal experimental data, which suggestive evidence for plausible pathways by which the risk of respiratory or cardiovascular mortality could be increased by ambient ozone.

Multiple time-series epidemiological studies explore the relationship between short-term ozone exposure and premature mortality. Most notably, a large multi-city study known as the National Morbidity, Mortality, and Air Pollution Study (NMMAPS) was designed to explore the association between several pollutants, including ozone, and daily mortality that focused on large cities across the US where levels of pollutants were varied (Samet et al., 2000). Two recently published studies based on the NMMAPS database that focus on the ozone/premature mortality relationship are Bell et al. (2004) (95 U.S. cities) and Huang et al. (2005) (19 U.S. cities). Another multi-city study by Schwartz (2005) examined the relationship between short-term ozone exposure and mortality in 14 U.S. cities.

In addition to these multi-city estimates, C-R functions for short-term ozone mortality can be derived from meta-analyses, which combine the results of several studies. Three meta-analyses were performed to obtain a summary estimate of ozone-related mortality risks and to attempt to describe heterogeneity in risk estimates (Ito et al., 2005; Levy et al., 2005; Bell et al., 2005). Each of these studies used different statistical techniques and datasets and examined statistical concerns, such as confounding, collinearity and possible interaction effects.[53]

In general, effect estimates from the meta-analyses are higher than the multi-city results. This could potentially be due to publication bias, as the meta-analyses relied solely on published studies, which could be more likely to contain statistically significant results. NMMAPS generally produces lower estimates than other epidemiological time-series studies, however, which could reflect specific methodological choices made by these investigators. Since these studies are associated with different strengths and limitations and no single study emerges as the most suitable to use as the basis for our primary estimate, we opted to use a pooled estimate, equally weighting the C-R functions from all six of these studies.

In addition to time-series epidemiological studies, a limited number of studies examine the cumulative effect of long-term exposure to ozone on mortality. One such recent study (Jerrett et al., 2009) used study population data from the ACS cohort study along with ozone monitoring data and reported a significant association between deaths from respiratory causes and long-term ozone exposure. In a recent review of the 812 Second Prospective Analysis methodology, the Council HES found the use of the Jerrett et al. estimate as the primary estimate premature at this time, due to a lack of corroboration from other cohort studies.[54]

Health Valuation Functions

In environmental benefit-cost analyses, the dollar value of an environmental benefit, such as improved health or avoidance of a case of illness, is the dollar amount necessary such that the person would be indifferent between experiencing the benefit and possessing the money. In most cases, the dollar amount required to compensate a person for exposure to an adverse effect is roughly the same as the dollar amount a person is willing to pay to avoid the effect. Therefore, in economic terms, the "willingness-to-pay" (WTP) is the appropriate measure of the value of avoiding an adverse effect. For example, the value of an avoided respiratory symptom would be a person's WTP to avoid that symptom.

For most goods, WTP can be observed by examining actual market transactions. For example, if a gallon of bottled drinking water sells for one dollar, it can be observed that at least those persons who choose to purchase that good are willing to pay at least one dollar for the water. For goods that are not exchanged in the market, such as most environmental goods, valuation is not so straightforward. Nevertheless, a value may be inferred from observed behavior, such as through estimation of the WTP for mortality risk reductions based on observed sales and prices of products that result in similar effects or risk reductions, (e.g., non-toxic cleaners or bike helmets). Alternatively, surveys may be used in an attempt to directly elicit WTP for an environmental improvement. Wherever possible in this analysis, we use estimates of mean WTP. In cases where WTP estimates are not available, we use the cost of treating or mitigating the effect as an alternative estimate.

For example, for the valuation of hospital admissions we use the avoided medical costs as an estimate of the value of avoiding the health effects causing the admission. These costs of illness (COI) estimates generally understate the true value of avoiding a health effect. They tend to reflect the direct expenditures related to treatment and not the utility an individual derives from improved health status or avoided health effect. We use a range of values for most environmental effects, to support the primary central estimate of net benefits. Table 5-4 summarizes the mean unit value estimates that we use in this analysis.

Valuation of Premature Mortality

Some forms of air pollution increase the probability that individuals will die prematurely. We use C-R functions for mortality that express the increase in mortality risk as cases of "excess premature mortality" per year. The benefit provided by air pollution reductions, however, is the avoidance of small increases in the risk of mortality. By summing individuals WTP to avoid small increases in risk over enough individuals, we can infer the value of a statistical premature death avoided.[55] For expository purposes, we express this valuation as "dollars per mortality avoided," or "value of a statistical life" (VSL), even though the actual valuation is of small changes in mortality risk experienced by a large number of people. The economic benefits associated with avoiding premature mortality were the largest category of monetized benefits in the First Prospective Analysis and continue to be the largest source of monetized benefits for this Second Prospective Analysis. Mortality benefits, however, are also the largest contributor to the range of uncertainty in monetized benefits.

Because avoided premature mortality benefits are such an important part of this study's results and findings, the remainder of this section provides an expanded discussion of some of the issues in valuing the avoidance of mortality risks from air pollution. We first discuss some characteristics of an "ideal" measure of the value of mortality risk reductions from air pollution, and then review several dimensions in which the current estimates fall short of the ideal measure for this study. For a more detailed discussion of the factors affecting the valuation of premature mortality see the Uncertainty section of this chapter and the *Uncertainty Analyses to Support the Second Section 812 Benefit-Cost Analysis of the Clean Air Act.*

Table 5-4. Unit Values for Economic Valuation of Health Endpoints (2006$).

HEALTH ENDPOINT	CENTRAL ESTIMATE OF VALUE PER STATISTICAL INCIDENCE		DERIVATION OF DISTRIBUTIONS OF ESTIMATES
	1990 INCOME LEVEL	2020 INCOME LEVEL	
Premature Mortality (Value of a Statistical Life)	$7,400,000	$8,900,000	Mean Value of Statistical Life (VSL) based 26 wage-risk and contingent valuation studies. A Weibull distribution, with a mean of $7.4 million (in 2006$), provided the best fit to the 26 estimates. Note that VSL represents the value of a small change in mortality risk aggregated over the affected population.
Chronic Bronchitis (CB)	$399,000	$490,000	The WTP to avoid a case of pollution-related CB is calculated as $WTP_x = WTP_{13} \cdot e^{-\beta(13-x)}$ where x is the severity of an average CB case, WTP 13 is the WTP for a severe case of CB, and β is the parameter distribution of WTP for an air pollution-relevant, average severity-level case of CB was generated by Monte Carlo methods, drawing from each of three distributions: (1) WTP to avoid a severe case of CB is assigned a 1/9 probability of being each of the first nine deciles of the distribution of WTP responses in Viscusi et al. (1991); (2) the severity of a pollution-related case of CB (relative to the case described in the Viscusi study) is assumed to have a triangular distribution, with the most likely value at severity level 6.5 and endpoints at 1.0 and 12.0; and (3) the constant in the elasticity of WTP with respect to severity is normally distributed with mean = 0.18 and standard deviation = 0.0669 (from Krupnick and Cropper (1992)). This process and the rationale for choosing it is described in detail in the Costs and Benefits of the Clean Air Act, 1990 to 2010 (EPA, 1999).
Nonfatal Myocardial Infarction (heart attack) 7% discount rate Age 0–24 Age 25–44 Age 45–54	 $ 84,171 $ 93,802 $ 98,366		No distributional information available. Age-specific cost-of-illness values reflect lost earnings and direct medical costs over a 5-year period following a nonfatal MI. Lost earnings estimates are based on Cropper and Krupnick (1990). Direct medical costs are based on simple average

Table 5-4. (Continued)

HEALTH ENDPOINT	CENTRAL ESTIMATE OF VALUE PER STATISTICAL INCIDENCE		DERIVATION OF DISTRIBUTIONS OF ESTIMATES
	1990 INCOME LEVEL	2020 INCOME LEVEL	
Age 55–65 Age 66 and over	$ 166,222 $ 84,171		of estimates from Russell et al. (1998) and Wittels et al. (1990). Lost earnings: Cropper and Krupnick (1990). Present discounted value of 5 years of lost earnings (2006$): age of onset: at 7%a 25–44 $9,631 45–54 $14,195 55–65 $82,051 Direct medical expenses: An average of (2006$): 1. Wittels et al. (1990) ($141,124—no discounting) 2. Russell et al. (1998), 5-year period ($28,787 at 3% discount rate; $ 27,217 at 7% discount rate)
Hospital Admissions			
All respiratory (ages 65+)	$23,711	$23,711	No distributions available. The COI point estimates (lost earnings plus direct medical costs) are based on ICD-9 code level information (e.g., average hospital care costs and average length of hospital stay) reported in Agency for Healthcare Research and Quality, 2000 (www.ahrq.gov). As noted in the text, no adjustments are made to cost of illness values for income growth.
All respiratory (ages 0–2)	$10,002	$10,002	
Chronic Obstructive Pulmonary Disease (COPD) (ages 65+)	$17,308	$17,308	
Asthma Admissions (ages <65)	$10,040	$10,040	
Pneumonia Admiss-ions (ages 65+)	$23,004	$23,004	
COPD, less asthma (ages 20–64)	$15,903	$15,903	
All Cardiovascular (ages 65+)	$27,319	$27,319	
All Cardiovascular (ages 20–64)	$29,364	$29,364	
Ischemic Heart Disease (ages 65+)	$33,357	$33,357	
Dysrhythmia (ages 65+)	$19,643	$19,643	
Congestive Heart Failure (ages 65+)	$19,619	$19,619	
Emergency Room Visits for Asthma	$369	$369	No distributional information available. Simple average of two unit COI values (2006$): (1) $401.62, from Smith et al. (1997) and (2) $336.03, from Stanford et al. (1999). As noted in the text, no adjust-ments are made to cost of illness values for income growth.
Respiratory Ailments Not Requiring Hospitalization			
Upper Respiratory Symptoms (URS)	$28.8	$30.7	Combinations of the three symptoms for which WTP estimates are available that closely match those listed by Pope et al.

Table 5-4. (Continued)

HEALTH ENDPOINT	CENTRAL ESTIMATE OF VALUE PER STATISTICAL INCIDENCE		DERIVATION OF DISTRIBUTIONS OF ESTIMATES
	1990 INCOME LEVEL	2020 INCOME LEVEL	
			result in seven different "symptom clusters," each describing a "type" of URS. A dollar value was derived for each type of URS, using mid-range estimates of WTP (IEc, 1994) to avoid each symptom in the cluster and assuming additivity of WTPs. In the absence of information surrounding the frequency with which each of the seven types of URS occurs within the URS symptom complex, we assumed a uniform distribution between $10.8 and $50.5 (2006$).
Lower Respiratory Symptoms (LRS)	$18	$19	Combinations of the four symptoms for which WTP estimates are available that closely match those listed by Schwartz et al. result in 11 different "symptom clusters," each describing a "type" of LRS. A dollar value was derived for each type of LRS, using mid-range estimates of WTP (IEc, 1994) to avoid each symptom in the cluster and assuming additivity of WTPs. The dollar value for LRS is the average of the dollar values for the 11 different types of LRS. In the absence of information surrounding the frequency with which each of the 11 types of LRS occurs within the LRS symptom complex, we assumed a uniform distribution between $8.1 and $28.6 (2006$).
Asthma Exacerbations	$50	$54	Asthma exacerbations are valued at $50 per incidence, based on the mean of average WTP estimates for the four severity definitions of a "bad asthma day," described in Rowe and Chestnut (1986). This study surveyed asthmatics to estimate WTP for avoidance of a "bad asthma day," as defined by the subjects. For purposes of valuation, an asthma exacerbation is assumed to be equivalent to a day in which asthma is moderate or worse as reported in the Rowe and Chestnut (1986) study. The value is assumed have a uniform distribution between $18.3 and $82.9 (2006$).
Acute Bronchitis	$416	$512	Assumes a 6-day episode, with the distribution of the daily value specified as uniform with the low and high values based on those recommended for related respiratory symptoms in Neumann et al. (1994). The low daily estimate of $20.5 (2006$) is the sum of the mid-range values recommended by IEc (1994) for two symptoms believed to be associated with acute bronchitis: coughing and chest

Table 5-4. (Continued)

HEALTH ENDPOINT	CENTRAL ESTIMATE OF VALUE PER STATISTICAL INCIDENCE		DERIVATION OF DISTRIBUTIONS OF ESTIMATES
	1990 INCOME LEVEL	2020 INCOME LEVEL	
			tightness. The high daily estimate was taken to be twice the value of a minor respiratory restricted activity day, or $118 (2006$). The low and high daily values are multiplied by six to get the 6-day episode values.
Work Loss Days (WLDs)	Variable (U.S. median = $149)		No distribution available. Point estimate is based on county-specific median annual wages divided by 50 (assuming 2 weeks of vacation) and then by 5—to get median daily wage. U.S. Year 2000 Census, compiled by Geolytics, Inc.
Minor Restricted Activity Days (MRADs)	$59	$64	Median WTP estimate to avoid one MRAD from Tolley et al. (1986). Distribution is assumed to be triangular with a minimum of $24 and a maximum of $94, with a most likely value of $59 (2006$). Range is based on assumption that value should exceed WTP for a single mild symptom (the highest estimate for a single symptom—for eye irritation—is $24) and be less than that for a WLD. The triangular distribution acknowledges that the actual value is likely to be closer to the point estimate than either extreme.
School Loss Days	$89	$89	No distribution available. Point estimate is based on (1) the probability that, if a school child stays home from school, a parent will have to stay home from work to care for the child, and (2) the value of the parent's lost productivity. Calculated using U.S. Bureau of Census data. School loss days, similar to cost of illness estimates for emergency room visits and hospital admissions, are not adjusted for changes in longitudinal income.

a These values are presented using a seven percent discount rate for this draft report, however these results will be presented using a five percent discount rate in the final report.

The health science literature on air pollution indicates that several human characteristics affect the degree to which mortality risk affects an individual. For example, some age groups appear to be more susceptible to air pollution than others (e.g., the elderly and children). Health status prior to exposure also affects susceptibility. At-risk individuals include those who have suffered strokes or are suffering from cardiovascular disease and angina (Rowlatt, et al. 1998). An ideal economic benefits estimate of mortality risk reduction would reflect these human characteristics, in addition to an individual's WTP to improve one's own chances of survival plus WTP to improve other individuals' survival rates.[56] The ideal measure would also take into account the specific nature of the risk reduction that is provided to individuals, as well as the context in which risk is reduced. To measure this value, it is important to assess how reductions in air pollution reduce the risk of dying from the time that reductions take effect onward, and how individuals value these changes. Each individual's survival curve, or the probability of surviving beyond a given age, should shift as a result of an environmental quality improvement. For example, changing the current probability of survival for an individual also shifts future probabilities of that individual's survival. This probability shift will differ across individuals because survival curves are dependent on such characteristics as age, health state, and the current age to which the individual is likely to survive.

A survival curve approach provides a theoretically preferred method for valuing the economic benefits of reduced risk of premature mortality associated with reducing air pollution, but the approach does not align well with current estimates of individual willingness to pay to avoid mortal risks. We have adopted the survival curve approach in the population simulation model that we use to generate estimates of life years lost and reduced life expectancy associated with air pollution, but implementing that approach requires that we use a national measure of the change in air pollution exposure, and also does not include a valuation component. As a result, the population simulation model results are not used for the primary results.

The Project Team also considered whether other evidence might support an adjustment to the VSL used in this study, particularly to account for the age of individuals affected. In general, studies of WTP to reduce mortality risk do not provide information on how VSL varies with life expectancy, but there are a few studies that attempt to assess the impact of age on VSL.[57] Some economic models in the theoretical literature suggest that VSL follows an inverted U, rising through middle age and falling at older ages, though this model is only partially supported by the relevant empirical evidence (Johansson 2002, Hammitt 2007). For example, revealed preference studies of the wage-risk literature support the inverted-U hypothesis (Aldy and Viscusi, 2007). These studies are limited, however, in that they necessarily include only employed workers and thereby exclude the elderly and those in poor health. Stated-preference studies, which can include a broader population, yield mixed results. Some suggest little or no effect of age on VSL and others suggest a modest decrease at older ages (Krupnick, 2007). Some studies, such as those by DeShazo (with Cameron, 2004), Chestnut (et al., 2004), and Alberini (et al., 2004) have found the effect of age on VSL to be statistically weak, suggesting a flatter relationship of VSL and age with a decline in VSL at much older ages. Consistent with Hammitt (2007), we conclude that there is insufficient evidence in the empirical VSL literature at this time to support an adjustment to the base VSL for the age of the affected population.

In sum, the economic valuation literature does not yet include good estimates of the value of this particular risk reduction commodity. As a result, in this study we value avoided premature mortality risk using the value of statistical life approach. As in the First Prospective Analysis, we use a mortality risk valuation estimate which is based on an analysis of 26 policy-relevant value-of-life studies (see Table 5-5). Five of the 26 studies are contingent valuation (CV) studies, which directly solicit WTP information from subjects; the remaining studies are wage-risk studies, which base WTP estimates on estimates of the additional compensation demanded in the labor market for riskier jobs.

We used the best estimate from each of the 26 studies to construct a distribution of mortality risk valuation estimates for the section 812 study. A Weibull distribution, with a mean of $7.4 million (in 2006$), provided the best fit to the 26 estimates.

Table 5-5. Summary of Mortality Valuation Estimates per Statistical Incidence of Premature Mortality (Millions of 2006$).

STUDY	TYPE OF ESTIMATE	VALUATION (MILLIONS 2006$)
Kneisner and Leeth (1991) (US)	Labor Market	$ 0.9
Smith and Gilbert (1984)	Labor Market	$ 1.1
Dillingham (1985)	Labor Market	$ 1.4
Butler (1983)	Labor Market	$ 1.7
Miller and Guria (1991)	Contingent Valuation	$ 1.9
Moore and Viscusi (1988a)	Labor Market	$ 3.9
Viscusi, Magat, and Huber (1991 b)	Contingent Valuation	$ 4.2
Gegax et al. (1985)	Contingent Valuation	$ 5.1
Marin and Psacharopoulos (1982)	Labor Market	$ 4.3
Kneisner and Leeth (1991) (Australia)	Labor Market	$ 5.1
Gerking, de Haan, and Schulze (1988)	Contingent Valuation	$ 5.2
Cousineau, Lacroix, and Girard (1988)	Labor Market	$ 5.6
Jones-Lee (1989)	Contingent Valuation	$ 5.9
Dillingham (1985)	Labor Market	$ 6.0
Viscusi (1978, 1979)	Labor Market	$ 6.3
R.S. Smith (1976)	Labor Market	$ 7.1
V.K. Smith (1976)	Labor Market	$ 7.2
Olson (1981)	Labor Market	$ 8.0
Viscusi (1981)	Labor Market	$ 10.0
R.S. Smith (1974)	Labor Market	$ 11.1
Moore and Viscusi (1988a)	Labor Market	$ 11.3
Kneisner and Leeth (1991) (Japan)	Labor Market	$ 11.7
Herzog and Schlottman (1987)	Labor Market	$ 14.0
Leigh and Folson (1984)	Labor Market	$ 15.0
Leigh (1987)	Labor Market	$ 16.0
Garen (1988)	Labor Market	$ 20.8

Source: Viscusi, 1992 and EPA analysis.

An additional uncertainty that is pertinent for this study's results is the potential bias in using estimates of VSL that correspond to small changes in risk for the relatively larger changes in mortality risk estimated in this study. As the results section below indicates, the large changes in $PM_{2.5}$ that represent the difference between the *with-CAAA* and *without-CAAA* scenarios by 2020 lead to a change in annual mortality risk of approximately 1 in one thousand for adults aged 25 and older, or 7 in ten thousand for all ages, which corresponds to a roughly ten percent change from the national baseline mortality risk of approximately 1 in one hundred.[58] This risk change is large compared to the mean mortality risk faced by subjects in the wage-risk studies that underlie our estimate of VSL – the mean risk for individual studies in our group of 26 varies from 4 in 10,000 to 5 in 100,000, although clearly some individuals in those samples face higher individual risks.[59] Economic theory suggests that individuals' incremental willingness to pay to reduce mortality risk declines with an increasing size of the risk increment, but the rate at which it declines is uncertain.[60] Estimates of differences in VSL across individuals in wage-risk study samples are also not informative, because they reflect variability in individuals' risk tolerance rather than differences in WTP across a population for varying increments of risk reduction. Further, it is not clear whether, in this context, the external risk imposed by air polluters on the exposed population implies that willingness-toaccept-compensation (WTAC) to forgo air quality improvement may be the more relevant measure. There is some theoretical work which suggests that, while valuation of a large risk increment may lead WTP estimates to be overestimated, it may lead WTAC estimates to be underestimated.[61] Although the Project Team remains concerned that there may be a potentially important disparity between the large increment of risk valued in this study and relatively smaller increments of risk valued in the underlying VSL literature, we conclude that the current literature does not provide a sufficient basis to make a quantitative adjustment to our base VSL values to account for this factor.

When valuing premature mortality for PM, we assume a lag between reduced PM exposure and the resulting reductions in incidences of premature mortality.[62] This lag does not affect the number of estimated incidences, but does alter the monetization of benefits. Because we value the "event" rather than the present risk, in this analysis we assume that the value of avoided future premature mortality should be discounted. The primary estimate reflects a 20-year distributed lag structure, which was recommended by the Council HES (2004). Under this scenario, 30 percent of the mortality reductions occur in the first year, 50 percent occur equally in years two through five, and the remaining 20 percent occur equally in years six through 20. Our valuation of avoided premature mortality applies a five percent discount rate to the lagged estimates over the periods 2000 to 2020, 2010 to 2030 and 2020 to 2040. We discount over the period between the initial PM exposure change (2000, 2010, or 2020) and the timing of the resulting change in incidence.

Health Effects Modeling Results

This section presents a summary of the differences in health effects resulting from improvements in air quality between the *with-CAAA* and the *without-CAAA* scenarios. Table 5-6 summarizes the CAAA-related avoided health effects in 2020 for each health endpoint included in the analysis and the associated monetary benefits. The mean estimate is presented as the primary central estimate, the 5th percentile observation is presented as the primary low estimate and the 95th percentile is presented as the primary high estimate.[63] In general, because the differences in air quality between the *with-* and *without-CAAA* scenarios are expected to increase from 1990 to 2020 and because population is also expected to increase during that time, the health benefits attributable to the CAAA are expected to increase consistently from 1990 to 2020. More detailed results can be found in *Health and Welfare Benefits Analyses to Support the Second Section 812 Benefit-Cost Analysis of the Clean Air Act*, February 2011.

Avoided Premature Mortality Estimates

Our analysis indicates that the benefit of avoided premature mortality risk reduction dominates the overall net benefit estimate. This is, in part, due to the high monetary value assigned to the avoidance of premature mortality relative to the unit value of other health endpoints. As described in detail in this chapter, there are also significant reductions in other short-term and chronic health effects and a substantial number of health benefits that we could not quantify or monetize. Mean results for all three target years are provided in Table 5-6, and the mean, primary low, and primary high estimates for 2020 are presented in Table 5-7.

As shown in Table 5-7, our primary central estimate implies that PM and ozone reductions due to the CAAA in 2020 will result in 230,000 avoided deaths, with a primary low and primary high bound on this estimate of 45,000 and 490,000 avoided deaths, respectively. These avoided deaths are valued at $1.8 trillion (2006$), with primary low and primary high bounds on this estimate of $170 billion to $5.5 trillion. To provide some context for these large values, we estimated the per capita risk change and monetized benefits. The estimated 230,000 avoided deaths in 2020 are equivalent to a total annual mortality risk reduction of 6.8×10^{-4} for the full estimated US population in 2020. With approximately 2.4 million estimated deaths in 2002, the avoided deaths in 2020 would increase total deaths by about 9.5 percent. The 230,000 avoided deaths are about 16 percent of the total mortality from the top four causes of death in the US in 2002: heart disease (over 600,000 deaths); cancer (over 550,000 deaths); stroke (over 130,000 deaths); and chronic lower respiratory disease (just less than 130,000 deaths). The monetized benefit per capita in 2020 is about $6,000, increasing from $2,700 in 2000 and $4,200 in 2010. Monetized benefits per household would be approximately $16,000 in 2020, increasing from $7,300 in 2000 and $11,000 in 2010.

It may also be worth noting that most of the changes in mortality risk we estimate occur in locations where both the *with-CAAA* and *without-CAAA* concentrations are above the lowest measured level (LML) in the underlying epidemiological studies. As noted above, standard EPA practice is to estimate PM-related mortality without applying an assumed concentration threshold, and the LML is itself not a threshold either. The LML approach summarizes the distribution of avoided PM mortality impacts according to the baseline $PM_{2.5}$ levels experienced by the population receiving the $PM_{2.5}$ mortality benefit. Unlike an assumed threshold, the LML is a characterization of the fraction of benefits that are more uncertain. In general, our confidence in the estimated PM mortality decreases as we consider air quality levels further below the LML in the two underlying PM-mortality epidemiological studies, Pope et al. (2002) and Laden et al. (2006).

Using the Pope et al. (2002) study, approximately 98 percent of the mortality impacts occur among populations with exposure to annual mean $PM_{2.5}$ levels at or above the LML of 7.5 $\mu g/m^3$. Using the Laden et al. (2006) study, approximately 91 percent of the mortality impacts occur at or above the LML of 10 $\mu g/m^3$. These analyses confirm that the great majority of the mortality benefits occur at or above the cohort study LMLs.

Avoided premature mortality is one of the more commonly cited results of benefits analyses for air pollution control. However, as noted in the valuation section of this chapter, a more accurate description of the benefit of clean air is a reduction in the risk of mortality for the exposed population over many years, which results in the extension of lives (sometimes referred to as "lives saved"). Other useful metrics of the benefit of cleaner air are the number of life years that are gained through the reduction of mortal risks, and the number of years of life expectancy gained on average throughout the population.

Table 5-6. Mean CAAA-Related Avoided Annual Incidence of Health Effects and Associated Monetary Valuation in 2000, 2010, and 2020.

ENDPOINT	POLLUTANT	INCIDENCE			VALUATION (MILLIONS 2006$)		
		2000	2010	2020	2000	2010	2020
Mortality							
Mortality – adults 30 and older	PM	110,000	160,000	230,000	$710,000	$1,200,000	$1,700,000
Mortality – infant	PM	160	230	280	$1,300	$1,900	$2,500
Mortality – all ages	Ozone	1,400	4,300	7,100	$10,000	$33,000	$55,000
Morbidity							
Chronic Bronchitis	PM	34,000	54,000	75,000	$14,000	$24,000	$36,000
Non-fatal Myocardial Infarction	PM	79,000	130,000	200,000	$8,100	$14,000	$21,000
Hospital Admissions Respiratory,	PM, Ozone	20,000	41,000	66,000	$290	$640	$1,100
Hospital Admissions, Cardiova-scular	PM	26,000	45,000	69,000	$760	$1,300	$2,000
Emergency Room Visits, Respiratory	PM, Ozone	58,000	86,000	120,000	$21	$32	$44
Acute Bronchitis	PM	96,000	130,000	180,000	$42	$61	$94
Lower Respiratory Symptoms	PM	1,200,000	1,700,000	2,300,000	$22	$30	$42
Upper Respiratory Symptoms	PM	980,000	1,400,000	2,000,000	$30	$42	$60
Asthma Exacerbation	PM	1,200,000	1,700,000	2,400,000	$61	$90	$130
Minor Restricted Activity Days	PM, Ozone	49,000,000	84,000,000	110,000,000	$2,900	$4,900	$6,700
Work Loss Days	PM	8,000,000	13,000,000	17,000,000	$1,300	$2,000	$2,700
School Loss Days	Ozone	1,200,000	3,200,000	5,400,000	$110	$290	$480
Outdoor Worker Productivity	Ozone	N/A	N/A	N/A	$30	$100	$170

Note: All incidence and valuation results are rounded to two significant figures. All estimates are annual estimates for individual target years of the analysis. Mortality valuation estimates reflect a delay in mortality incidence from the time of the exposure change in the target year, reflecting application of a 20-year distributed cessation lag as described in the text and a 5 percent discount rate.

Table 5-7. Caaa-Related Avoided Annual Incidence of Health EFFECTS And Associated Monetary Valuation in 2020.

ENDPOINT	POLLUTANT	INCIDENCE			VALUATION (MILLIONS 2006$)		
		5TH %ILE	MEAN	95TH %ILE	5TH %ILE	MEAN	95TH %ILE
Mortality							
Mortality[1]	PM, Ozone	45,000	230,000	490,000	$170,000	$1,800,000	$5,500,000
Morbidity							
Chronic Bronchitis	PM	12,000	75,000	130,000	$3,100	$36,000	$130,000
Non-fatal Myocardial Infarction	PM	80,000	200,000	300,000	$6,200	$21,000	$48,000
Hospital Admissions, Respiratory	PM, Ozone	24,000	66,000	110,000	$320	$1,100	$1,800
Hospital Admissions, Cardiovascular	PM	52,000	69,000	84,000	$1,400	$2,000	$2,600
Emergency Room Visits, Respiratory	PM, Ozone	64,000	120,000	180,000	$22	$44	$69
Acute Bronchitis	PM	-7,000	180,000	340,000	-$4	$94	$220
Lower Respiratory Symptoms	PM	1,200,000	2,300,000	3,300,000	$18	$42	$76
Upper Respiratory Symptoms	PM	620,000	2,000,000	3,300,000	$17	$60	$130
Asthma Exacerbation	PM	270,000	2,400,000	6,700,000	$15	$130	$390
Minor Restricted Activity Days	PM, Ozone	91,000,000	110,000,000	140,000,000	$3,800	$6,700	$10,000
Work Loss Days	PM	15,000,000	17,000,000	19,000,000	$2,300	$2,700	$3,000
School Loss Days	Ozone	2,200,000	5,400,000	8,600,000	$190	$480	$770
Outdoor Worker Productivity	Ozone	N/A	N/A	N/A	$170	$170	$170

Notes: [1] Includes adult and infant mortality for PM and all ages for ozone.

All incidence and valuation results are rounded to two significant figures. Mortality valuation estimates reflect a delay in mortality incidence from the time of the exposure change in the target year, reflecting application of a 20-year distributed cessation lag as described in the text and a 5 percent discount rate.

In addition, we tracked the impact of these effects, held constant at the 2020 levels, for an additional 30 years, through 2050. Running the simulation beyond 2020 allows us to estimate the full effect of changes that begin in 2020, which because of the cessation lag are not fully realized until many years after the end of the study period. Comparing the estimated population in each age cohort across the two scenarios allows us to estimate gains in life-years (i.e., one additional person in a cohort for one year yields a life year gained), and summing across cohorts and years yields cumulative estimates. In addition, analysis of the changes in mortality risk among cohorts older than a specific age yields estimates of life expectancy gains at specific ages.[64]

The results of these calculations are presented in Table 5-8 below, and provide further evidence of the substantial benefits of CAAA during and after the 1990-2020 period. The first panel of the table provides estimates of life-years gained for 2020 and 2040 – these are estimates of the life-years gained only in that year of the simulation, but reflect the cumulative effect of mortality risk reductions in prior years. The next panel provides estimates of cumulative life years gained overall all years since 1990, first for the 1990- 2020 period, and then for the 1990-2040 period, inclusive.

Table 5-8. Life Years Gained and Life Expectancy Gain Estimates from the Population Simulation Model.

AGE COHORT		LIFE-YEARS GAINED IN SPECIFIC YEARS (ANNUAL)		CUMULATIVE LIFE YEARS GAINED THROUGH TARGET YEAR		LIFE EXPECTANCY GAINS (YEARS)		
START AGE	END AGE	2020	2040	2020	2040	2010	2020	2040
30	39	17,000	18,000	260,000	620,000	0.65	0.87	0.91
40	49	60,000	71,000	910,000	2,300,000	0.63	0.84	0.88
50	59	150,000	180,000	2,000,000	5,400,000	0.59	0.79	0.84
60	69	330,000	380,000	3,500,000	11,000,000	0.53	0.71	0.76
70	79	470,000	840,000	5,000,000	20,000,000	0.44	0.59	0.64
80	89	470,000	1,200,000	6,000,000	23,000,000	0.32	0.43	0.48
90	99	320,000	800,000	3,600,000	14,000,000	0.19	0.25	0.27
100+		60,000	200,000	490,000	3,100,000	0	0	0
Total		1,900,000	3,800,000	22,000,000	80,000,000			

Note: Column entries to not add to totals due to rounding. Life expectancy results are incremental period conditional life expectancy gains at the start age of the cohort.

As expected, life-years gained are largest in the older cohorts, particularly cohorts 60 years and older, and they increase over time as the effect of mortality risk reduction in successive years increases survival rates among all individuals age 30 and over. By 2020, the cumulative effects indicate 22 million life-years are gained from the air pollution mortality risk reduction.

The last panel provides the life expectancy results. As early as 2010, the CAAA increased life expectancy at 30 years by 0.65 years, with somewhat smaller gains among older cohorts. By 2040, the full effect of the CAAA on life expectancy is realized, with a total gain in life expectancy of almost one year at age 30 across the entire US population.

Non-Fatal Health Impacts

We report non-fatal health effects estimates in a similar manner to estimates of premature mortality – as a range of estimates for each quantified health endpoint, with the range dependent on the quantified uncertainties in the underlying C-R functions. The range of results for 2020 is characterized in Table 5-6 with 5th percentile, mean, and 95th percentile estimates which correspond to the primary low, central, and high estimates. All estimates are expressed as new cases avoided in 2020, with the following exceptions. Hospital admissions reflect admissions for a range of respiratory and cardiovascular diseases and these results, along with emergency room visits for respiratory disease, do not necessarily represent the avoidance of new cases of disease (i.e., air pollution may simply exacerbate an existing condition, resulting in an emergency room visit or hospital admission). Further, each admission is only counted once, regardless of the length of stay in the hospital. Minor restricted activity days, school loss days, and work loss days are expressed in terms of person-days. For instance, one "case" of a school loss day represents one person out of school for one day.

Avoided Health Effects of Air Toxics

The prior discussion focuses on the effects of the 1990 CAAA on particulate matter and ozone health effects, but the Amendments also address the control of air toxics or hazardous air pollutants (HAPs). HAPs are pollutants regulated under Title III of the CAAA that can cause adverse effects to human health and ecological resources. The Amendments establish a list of HAPs to be regulated, require EPA to establish air toxic emissions standards based on Maximum Achievable Control Technology (MACT) standards, and include a provision that requires EPA to establish more stringent air toxics standards if MACT controls do not sufficiently protect the public health against residual risks. Control of air toxics is expected to result both from these changes and from incidental control due to changes in criteria pollutant programs, such as controls on volatile organic compounds (VOCs) necessary to achieve the NAAQS for ambient tropospheric ozone.

Both the Retrospective analysis and the First Prospective analysis omitted a quantitative estimation of the benefits of reduced concentrations of air toxics, citing gaps in the toxicological database, difficulty in designing population-based epidemiological studies with sufficient power to detect health effects, limited ambient and personal exposure monitoring data, limited data to estimate exposures in some critical microenvironments, and insufficient economic research to support valuation of the types of health impacts often associated with exposure to individual air toxics. Based on a recommendation by the Council, EPA developed a case study of the benefits of CAAA controls on benzene emissions in the Houston area (USEPA, 200 1).[65] The purpose of the case study was to demonstrate a methodology that could be used to generate human health benefits from CAAA controls on a single HAP in an urban setting, while highlighting key limitations and uncertainties in the process. In addition,

EPA hoped to gain insight into the use of the case study methodology for characterizing benefits nationwide. The case study was not intended, however, to provide a comprehensive assessment of the benefits of benzene reductions due to the CAAA.

The case study involved calculating the reduction in the annual number of cases of leukemia due to reductions in benzene levels resulting from the 1990 CAAA through the year 2020 in the Houston metropolitan area. Benzene was selected for the case study due to the availability of human epidemiological studies linking its exposure with adverse health effects. The case study focused on Houston because of the presence of significant large benzene emitting sources, such as petroleum refineries, as well as sources more typical of other urban areas, such as gasoline refueling stations.

We conducted the case study using the same five steps used in the main 812 criteria pollutant analysis:

1. **Scenario Development:** We assessed benefits from the reduction in benzene concentrations between a *without-CAAA* scenario, which essentially freezes federal, state, and local air pollution controls at the levels of stringency and effectiveness that existed in 1990, and a *with-CAAA* scenario, which assumes that all federal, state, and local rules promulgated pursuant to, or in support of, the 1990 CAAA were implemented.
2. **Emissions Estimation:** We estimated benzene emissions in the Houston area under both the *with-CAAA* and *without-CAAA* scenarios by extrapolating data based on expected growth in emissions-generating activities over time, adjusted for the impact of future year control assumptions under each scenario.
3. **Air Quality and Exposure Modeling:** We then applied EPA's American Meteorological Society/Regulatory Model (AERMOD) dispersion modeling system (USEPA, 2004) to convert emissions estimates to ambient benzene concentrations at the Census block group level. The AERMOD output was then run through EPA's Hazardous Air Pollutant Exposure Model, Version 6 (HAPEM6; ICF International, 2007) to generate benzene exposure concentrations for the study population at the Census tract level, which reflect average benzene concentrations likely experienced by the study population as they carry out their daily activities.
4. **Health Effects Modeling:** We next estimated avoided cases of leukemia using a life-table based risk assessment model. The life-table model assessed age- specific risks at the Census tract level, based on county-level background rates of leukemia, age-specific benzene exposure data from HAPEM6 and an epidemiological dose-response function derived from a study of occupational benzene exposures (Crump, 1994).[66] The model yielded annual age-specific Census tract-level avoided cases of leukemia (fatal and non-fatal) for each target year. We also estimated the number of cases expected to occur after the end of the study period resulting from CAAA-related benzene changes within the study period, due to lagging effects of these changes on leukemia risks.
5. **Valuation:** We then applied valuation methods from the current economic literature to assign monetary value to the avoided leukemia cases. This included valuing fatal cancers using the VSL estimate used in the primary 812 analysis (i.e., the Weibull distribution based on 26 studies) with an adjustment for medical costs associated with the period of cancer illness leading up to death (i.e., "pre-mortality

morbidity").[67] We valued non-fatal cancers using two bounding estimates, a WTP value for chronic bronchitis and one from a health risk tradeoff study that provided a value for avoiding a case of non-fatal lymphoma.[68]

Table 5-9 presents our primary estimate for avoided fatal and non-fatal cases of leukemia due to CAAA-related changes in ambient benzene levels in the Houston area. It includes the number of expected annual cases avoided in each study year as well as the total cumulative avoided cases throughout the study period and the total cumulative avoided cases expected to occur after 2020, due to changes in benzene occurring within the study period. It also shows the monetary value (the 1990 net present value (NPV), using a five percent discount rate) of these avoided leukemia cases.

Table 5-9. Total Annual Benefits for Each Study Year from CAAA-Related Changes in Benzene Exposure in the Houston Area.

	ANNUAL AVOIDED CASES OF LEUKEMIA			TOTAL MONETARY BENEFITS, 1990 TO 2010 (1990 NPV, MILLIONS OF 2006$, 5% DISCOUNT RATE)		
	AVOIDED FATAL CASES	AVOIDED NON-FATAL CASES	TOTAL AVOIDED CASES	BENEFITS FROM FATAL CASES OF LEUKEMIA	BENEFITS FROM NON-FATAL CASES OF LEUKEMIA	TOTAL BENEFITS
Results by Study Year						
2000	0.03	0.02	**0.05**	$0.12	$0.01 – 0.06	**$0.13 – 0.18**
2010	0.09	0.07	**0.2**	$0.27	$0.01 – 0.13	**$0.28 – 0.40**
2020	0.2	0.1	**0.3**	$0.31	$0.01 – 0.15	**$0.32 – 0.46**
Cumulative Results						
Cumulative Cases Occurring within the Study Period	**2**	**2**	**4**	**$6.7**	**$0.32 – 3.3**	**$7.0 – 10**
Additional Cumulative Cases Occurring after 2020*	**1**	**1**	**2**	**$1.8**	**$0.08 – 0.8**	**$1.9 – 2.6**
Total Cumulative Cases	**3**	**3**	**6**	**$8.5**	**$0.40 – 4.1**	**$8.9 – 13**

* Note: These avoided cases are due to changes in benzene exposure that took place within the study period. However, the cases occurred after 2020 due to lagging effects of these changes on leukemia risks, as described in the text.

Our results indicate that by the year 2020, the change in benzene-related population risk due to the 1990 CAAA programs would be equivalent to a total of four cases of leukemia in the Houston area, with three of those occurring in Harris County, the most densely populated county included in the analysis. We estimated two of the four cases to be fatal and two to be non-fatal. Our primary central estimate of total benefits due to CAAArelated reductions in benzene is $8.9 to 13 million (in 2006$), $8.5 million of which is due to fatal cases of leukemia, and $0.4 to 4.1 million of which is due to non-fatal cases.

In addition to the leukemia analysis, we evaluated the numbers of individuals likely to be exposed to benzene at levels exceeding EPA's chronic reference concentration (RfC) for benzene, which is based on changes in white blood cell counts, under the *with-CAAA* and *without-CAAA* scenarios. We found no individuals exposed to benzene at concentrations exceeding the RfC in either the *with-CAAA* or *without-CAAA* scenario. We also conducted illustrative analyses of exposure and risk reductions to highly exposed subpopulations in the study area, and found potentially significant individual risk reductions due to the CAAA for individuals in these groups. For instance, a back-of-the envelope calculation of residents living in homes with attached garages, who are expected to have higher benzene exposures, suggests that adding attached garage-related benefits to our primary estimate could result in an approximate doubling of our primary estimate.

In summary, this case study demonstrates that the 1990 CAAA controls on benzene emissions are expected to result in reductions in the incidence of leukemia in the greater Houston area over the period 1990 to 2020. The case study does have some limitations, including possible underestimation of benzene emissions from large point sources (e.g., refineries), possible exclusion of unquantifiable adverse health effects of benzene (e.g., Hodgkin's and non-Hodgkin's Lymphoma), and exclusion of new programs established after the case study (e.g., Mobile Source Air Toxics Rule). However, it successfully demonstrates a methodology that can serve as a useful tool in EPA's evolving HAP benefits assessment strategy. It can provide a comprehensive assessment of the impact of benzene controls from multiple CAAA Titles on cancer incidence in an urban population, using a combination of national and local data to conduct urban-scale modeling of air quality and health impacts. Further, the life-table model allows for more careful assessment of risk changes over time at the Census tract level, incorporating local, age- specific baseline incidence data with age-specific exposure data and information on the lag between exposure changes and risk reductions.

Determining where this approach might fit within EPA's HAP benefits assessment strategy will require additional analysis and evaluation to determine the added value of the detailed, urban-scale approach, as well as the potential pool of HAPs suitable for assessment via the damage-function approach for cancer and/or non-cancer effects.

Comparison of Health Effects Modeling with First Prospective Analysis

Differences in Methodology

In comparison with the First Prospective 812 Analysis, the Second Prospective includes a number of refinements and improvements in health benefits estimation methods.

- *Targeted Criteria Pollutant Analysis:* The Second Prospective excludes benefits of CAAA-related reductions in carbon monoxide, nitrogen oxides, and sulfur dioxide, which were included in the First Prospective, in an effort to streamline the quantitative analysis to focus on the two criteria pollutants that yield the greatest benefits – $PM_{2.5}$ and ozone.

- *New Cessation Lag Structure for PM Mortality:* The Second Prospective relies on the use of a 20-year distributed lag structure assumption for the cessation lag between changes in PM exposure and resulting changes in premature mortality. This estimate represents a shift from the First Prospective, which applied a 5-year distributed lag based on smoking cessation literature. The 20-year distributed lag is based on recommendations from the Council HES, is derived from air pollution literature and attempts to more closely reflect the disease processes that occur from PM exposure.[69]
- *New C-R Function for PM Mortality:* The First Prospective relied upon a C-R function derived from the most recently published ACS cohort study at the time (Pope et al., 1995). Since this time, additional follow-up has occurred for both the ACS and Six Cities cohort studies. In addition, new evidence has emerged on the ACS study results that suggest that this estimate is potentially underestimated. Our new primary C-R function mean is based on the follow-up literature, specifically the Pope et al. (2002) update of the ACS cohort and the Laden et al. (2006) update of the Six Cities cohort. Our new C-R function also reflects the results of an expert elicitation study, which allowed experts to incorporate multiple sources of uncertainty in the C-R function and to adjust the C-R function estimates to account for known biases.
- *Ozone Mortality Benefits Estimates:* The Second Prospective includes ozone- related premature mortality. This additional endpoint, which was not included in the First Prospective, was added because of advances that have occurred in the epidemiological literature that provide consistent evidence for this health endpoint.[70]
- *New Health Benefits Modeling Program:* The Second Prospective relies on EPA's BenMAP health benefits modeling program. Key advantages of the updated model are ease of use, allowing us to more readily perform multiple sensitivity tests; updated population and baseline incidence estimates; new C-R function options; and the ability to perform integrated exposure analysis using the eVNA method described earlier.
- *Air Toxics Case Study:* The Second Prospective includes the results of a case study demonstrating a methodology for assessing health benefits from a single hazardous air pollutant.

Differences in Health Effects Modeling Results

The health effects estimates for the Second Prospective are much larger than the estimates EPA developed for the First Prospective. The 2020 estimates are new to the Second Prospective, but the comparable mean estimate of health benefits in 2000 and 2010 for the First Prospective were $71 billion in 2000 and $110 billion in 2010, in 1990$[71] - if updated to 2006$, these estimates would be $110 billion in 2000 and $170 billion in 2010. The Second Prospective results are larger by roughly a factor of 10. There are four key reasons we have identified for the increase in benefits:

1. ***Scenario differences***: The *with-CAAA* scenario, especially for the 2010 target year, includes new rules with substantial additional pollutant reductions that were not included in the comparable First Prospective scenario, such as the Clean Air Interstate Rule (CAIR).

2. ***Improved air quality models***: The First Prospective relied on the Regional Acid Deposition Model/Regional Particulate Model (RADM/RPM) for PM and deposition estimates in the eastern U.S., the Regulatory Modeling System for Aerosols and Acid Deposition (REMSAD) for PM estimates in the western U.S., and the Urban Airshed Model (versions V and IV) at various regional and urban scales to generate ozone estimates. The Second Prospective relies on the integrated CMAQ modeling tool, which reflects substantial improvements in air quality modeling, provides more comprehensive spatial coverage, and achieves improved model performance.
3. ***Better, more comprehensive exposure estimates***: The First Prospective relied on first generation exposure extrapolation tools to generate monitor-adjusted exposure estimates away from monitors. Since then, the monitor network, availability of speciated data, and the performance of speciated exposure estimation tools have improved substantially.
4. ***Updated dose-response estimates***: Since 1999, some concentration response functions have been updated, most notably the PM-premature mortality C/R function, whose central estimate of the mortality impact of fine PM has nearly doubled. In addition, health effects research has addressed endpoints that were not covered in the First Prospective, including premature mortality associated with ozone exposure.

Although the Agency has not yet conducted a rigorous quantitative analysis to assess the impact of these methodology and data improvements, and the differences in study design between the first and Second Prospective made such an analysis difficult to perform, the impact of most of these factors is to increase the estimates of benefits, in some cases very substantially.

Uncertainty in Health Benefits Estimates

A number of important assumptions and uncertainties in the health benefits analysis may influence the estimate of monetary benefits presented in this study. In this section of the chapter, we first discuss several quantitative sensitivity analyses undertaken to characterize the impact of key assumptions on the ultimate health benefits estimates. We then conclude with a qualitative discussion of the impact of both quantified and unquantified sources of uncertainty.

Quantitative Sensitivity Tests

We performed three quantitative sensitivity tests to estimate the impact of alternate assumptions on our overall benefits estimates due to avoided premature mortality, the largest contributor to our overall health benefits estimates. The three focal areas for sensitivity analysis were: (1) the C-R function estimate; (2) the PM/mortality cessation lag structure; and (3) the mortality valuation estimate (including both the VSL and the discount rate). These are influential assumptions in our analysis and those for which plausible alternative quantitative estimates are available. Table 5-10 below provides the results of these sensitivity analyses.

Concentration-Response Function

Our monetized estimate of the benefits of reducing premature mortality from CAAA-related pollution reductions is based on a single primary estimate C-R function for each of the criteria pollutants included in our analysis, $PM_{2.5}$ and ozone. This selection is associated with uncertainty related to potential across-study variation. That is, different published studies of the same pollutant/health effect relationship often do not report identical findings; in some instances, the differences are substantial. These differences can arise from differences in factors such as study design, random sampling for subject populations, or modeling choices, such as inclusion of potential confounders.

In order to estimate the effect of across-study variation on our CAAA-related mortality benefits from reductions in $PM_{2.5}$ and ozone, we performed a sensitivity analysis on the C-R functions selected. For $PM_{2.5,}$ our primary estimate is based on a Weibull distribution of C-R coefficients with a mean of 1.06 percent decrease in annual all-cause mortality per 1 jtg/m^3 and an interquartile range bracketed by the Pope et al. (2002) ACS estimate (0.55 percent) on the low end and the Six Cities Laden et al. (2006) extended follow-up estimate (1.5 percent) at the high end. We conducted a sensitivity analysis by first substituting the primary C-R distribution with alternative C-R functions, one based on the Pope et al. (2002) ACS study, one based on the Laden et al. (2006) Six Cities cohort study as well as the C-R distributions provided by each of the 12 experts included in the PM/mortality expert elicitation study.

For ozone, our primary estimate consists of a pooled estimate of six studies, three multi-city studies (Schwartz, 2005; Bell et al., 2004; Huang et al., 2005) and three meta-analyses (Ito et al., 2005; Levy et al., 2005; Bell et al., 2005). We conducted a sensitivity analysis by substitute this primary C-R function with the C-R functions reported in each of these six individual studies, and separately for the Jerrett et al. (2009) cohort study.

As shown in Table 5-10, substituting alternate PM C-R functions results in total mortality benefits estimates that range from between 81 percent lower up to 78 percent higher than the primary estimate. Substituting alternative ozone C-R function does not affect the total mortality benefits estimate, since ozone does not contribute significantly to this estimate. However, the C-R function selection does affect the ozone mortality estimates, ranging from 63 percent lower up to 66 percent higher than the primary estimate for ozone mortality incidence. As expected, the Jerrett et al. study yields estimates higher than the primary pooled estimate. Cohort studies measure the effects of cumulative exposure and so should reasonably yield higher estimates than the comparably parameterized time- series study - but within the range of underlying six studies, albeit at the high end of that range.

PM/Mortality Cessation Lag

The timing of the cessation lag between PM exposure and mortality remains uncertain. Our primary monetized estimate of PM/mortality benefits assumes a 20-year distributed lag (30 percent of the mortality reductions occur in the first year, 50 percent occur equally in years two through five, and the remaining 20 percent occur equally in years six through 20). We tested the sensitivity of this assumption by calculating monetized mortality benefits based on alternative cessation lag structures. We selected two alternative lag structures – a 5-year distributed lag (which was employed in the First Prospective) and a smooth function (which assumes an exponential decay model and is based on an analysis by Roosli et al., 2005; see Chapter 6 of *Uncertainty Analyses to Support the Second Section 812 Benefit-Cost Analysis of the Clean Air Act* for further details). We also calculated benefits assuming no cessation lag. Application of alternative cessation lag structures had a smaller impact on the benefits estimates than the C-R function, resulting in benefits estimates that range from 22 percent lower up to 16 percent higher than the primary estimate.

Mortality Valuation

We apply a VSL value to reductions in premature mortality based on a Weibull distribution of 26 study estimates. The literature on VSL is extensive, and studies have measured VSL using different methodological approaches (e.g., revealed versus stated preference) on a variety of study populations (e.g., workers versus a general population sample) in a variety of different risk contexts (e.g., fatal workplace accidents versus mortality risk from disease). In addition, several meta-analyses of the literature have been conducted in an attempt to synthesize the literature. As a result, there are many options for alternative VSL estimates. We selected several alternative VSL estimates derived from the literature for sensitivity testing, including two estimates from a meta- analysis by Viscusi and Aldy (2003), an estimate used in past EPA regulatory analyses in the form of a normal distribution, and an estimate from a wage-risk study by Viscusi (2004). VSL did not affect the benefits results to the same degree as the C-R function, with alternative monetized benefits ranging from 21 percent lower to approximately equivalent to our primary estimate.

Table 5-10. Results of Quantitative Sensitivity Tests.

FACTOR	STRATEGY FOR SENSITIVITY ANALYSIS	RANGE OF PERCENT CHANGES FROM MEAN PRIMARY MORTALITY BENEFITS ESTIMATE[1]
PM C-R Function	Alternative C-R functions – two from empirical literature (Pope et al., 2002 and Laden et al., 2006) and 12 subjective estimates from the expert elicitation study	-81% to 78%, Based on most extreme estimates from PM expert elicitation study. Rest of alter-natives range from -41% to 40%.
Ozone C-R Function	Alternative C-R functions – three from multi-city studies, three meta-analyses, and the Jerrett et al. (2009) cohort long-term exposure study	0% for total mortality benefits. -63% to 66% For ozone-related mortality.
PM/Mortality Cessation Lag	Alternative lag structures – one step function and one smooth function (based on an exponential decay function)	-22% to 16%
VSL	Alternative VSL estimates	-21% to 0%
Discount Rate	Alternative discount rates	-6% to 6%

[1] All values in the table represent the percent change from the mean primary estimate. Percent change estimates to not vary by target year.

Table 5-11. Key Uncertainties Associated with Human Health Effects Modeling.

POTENTIAL SOURCE OF ERROR	DIRECTION OF POTENTIAL BIAS FOR NET BENEFITS ESTIMATE	MAGNITUDE OF IMPACT ON NET BENFITS ESTIMATE	DEGREE OF CONFIDENCE
UNCERTAINTIES RELATED TO PREMATURE MORTALITY BENEFITS ESTIMATES			
Analysis assumes a causal relationship between PM exposure and premature mortality based on strong epidemiological evidence of a PM/mortality association. However, epidemiological evidence alone cannot establish this causal link.	Overestimate	Potentially major. PM/mortality effects are the largest contributor to the net benefits estimate. If the PM/mortality relationship is not causal, it would lead to a significant overestimation of net benefits.	High. The assumption of causality is suggested by the epidemiologic and toxicological evidence and is consistent with current practice in the development of a best estimate of air pollution-related health benefits. At this time, we can identify no basis to support a conclusion that such an assumption results in a known or suspected overestimation bias.
Analysis assumes a causal relationship between ozone exposure and premature mortality based on strong epidemiological and experimental evidence of an ozone/mortality association.	Overestimate	Probably minor. Ozone mortality effects are a large contributor to the net benefits estimate, but total monetized ozone mortality benefits remain less than five percent of total net benefits. If the ozone mortality relationship is not causal, it would lead to an overestimation of net benefits.	Medium. Several epidemiological studies provide strong evidence for associations between ozone and mortality. This data is supported by human and animal experimental studies that provide suggestive evidence for plausible mechanisms. Overall, the evidence is highly suggestive, but additional research is needed to more fully establish underlying mechanisms.

Table 5-11. (Continued)

POTENTIAL SOURCE OF ERROR	DIRECTION OF POTENTIAL BIAS FOR NET BENEFITS ESTIMATE	MAGNITUDE OF IMPACT ON NET BENFITS ESTIMATE	DEGREE OF CONFIDENCE
It is possible that the PM/mortality relationship is modified by socioeconomic status (SES).	Unable to determine based on current information. Consider-ation of both the Pope and Laden studies avoids the possible underestimation effect from the ACS cohort, owing to the demographics of that study population, and the possible overestimation bias associated with the more limited geographic scope of the Six Cities cohort.	Potentially major. Sensitivity analyses reported in this chapter indicate the high sensitivity of benefits results to the choice of the PM/mortality C/R function.	Medium. Studies have found effect modify-cation of the PM/mortality effect by SES, as assessed through edu-cation attainment (Krewski et al., 2000). However, this effect is likely to affect only the Pope et al. estimate. Our inclusion of both the Pope et al. and Laden et al. (which does includes a more diverse population) helps account for the possible significance of this uncertainty.
Exposure misclassification due to reliance on ambient monitoring data to estimate $PM_{2.5}$ exposures rather than measuring personal exposures.	Underestimate. Concentrations measured at central site monitors may not accurately reflect exposure experienced by the population due to variation in ambient concentrations over space within a geographic area, incomplete penetration of ambient pollution into homes and workplaces, patterns of population activity and indoor sources that can contribute significantly to individual $PM_{2.5}$ exposures. Reducing exposure	Potentially major. Recent analyses reported in Krewski et al. (2009) demonstrate the relatively significant effect that this source of uncertainty can have on effect estimates.	High. The results from Krewski et al. (2009) and Jerrett et al. (2005) suggest that exposure error may underestimate effect estimates (PM ISA).

Table 5-11. (Continued)

POTENTIAL SOURCE OF ERROR	DIRECTION OF POTENTIAL BIAS FOR NET BENEFITS ESTIMATE	MAGNITUDE OF IMPACT ON NET BENFITS ESTIMATE	DEGREE OF CONFIDENCE
	error can result in stronger associations between pollutants and health effects than generally observed in studies having less exposure detail.		
Exclusion of C-R functions from short-term exposure studies in PM mortality calculations.	Underestimate	Potentially major. PM/mortality is the top contributor to the net benefits estimate. If short-term functions contribute substantially to the overall PM-related mortality estimate, then the net benefits could be underestimated.	Medium. Long-term PM exposure studies likely capture a large part of the impact of short-term peak exposure on mortality; however, the extent of overlap between the two study types is unclear.
Assumption that PM-related mortality occurs over a period of 20 years following the critical PM exposure. Analysis assumes that 30% of mortality reductions in the first year, 50% over years 2 to 5, and 20% over the years 6 to 20 after the reduction in $PM_{2.5}$	Unable to determine based on current information	Potentially major. PM/mortality is the largest contributor to monetary benefits. Our quantitative sensitivity analysis indicated that alternative plausible cessation lag structures could alter the benefits estimate between 23% lower to 16% higher than the primary estimate.	Medium. Recent epidemiological studies (e.g., Schwartz, 2008) have shown that the majority of the risk occurs within 2 years of reduced exposure. However, our default lag assumes 43% of mortality reductions would occur within the first 2 years. The evidence directly informing the cessation lag structure is some-what limited, but the current lag is supported by the Council HES.
Assumption of a linear, no-threshold model for PM and ozone mortality	Overestimate	Probably minor. Although consideration for alternative model forms (Krewski et al., 2009) does	High. The current scientific literature does not support a population-

Table 5-11. (Continued)

POTENTIAL SOURCE OF ERROR	DIRECTION OF POTENTIAL BIAS FOR NET BENEFITS ESTIMATE	MAGNITUDE OF IMPACT ON NET BENFITS ESTIMATE	DEGREE OF CONFIDENCE
		suggest that different models can impact risk estimates to a certain extent, generally this appears to be a moderate source of overall uncertainty.	based threshold, which consis-tently shows effects down to the lowest measureable levels. If a threshold does exist, it is likely below the range of concen-trations of regulatory interest.
Mortality health impact did not include pollutants other than PM or ozone.	Unable to determine based on current information	Probably minor. If other criteria Pollutants correlated with PM contribute to mortality, that effect may be captured in the PM estimate. This uncertainty does make it difficult to disaggregate avoided mortality benefits by pollutant.	High. PM and ozone are the two pollutants most strongly linked to mortality in the epidemiological literature. It is likely that we've captured the majority of mortality benefits due to criteria pollutants in our analysis.
Pooling with equal weights of ozone mortality incidence estimates to present a primary estimate.	Unable to determine based on current information	Probably minor. Pooling with equal weights provides a central estimate of ozone mortality benefits, but it is not clear that the six ozone mortality incidence studies should be combi-ned in this manner. Relying on a particular single study or another combination of studies may result in significantly different estimated benefits from ozone reductions. However, ozone-related avoided mortality benefits are a minor con-tributor to total monetized benefits.	Medium. All six studies are associated with different strengths and limitation. No single study has emerged as solely suitable to support a primary estimate. Therefore, a pooled estimate provides a central estimate of the available literature.

Table 5-11. (Continued)

POTENTIAL SOURCE OF ERROR	DIRECTION OF POTENTIAL BIAS FOR NET BENEFITS ESTIMATE	MAGNITUDE OF IMPACT ON NET BENFITS ESTIMATE	DEGREE OF CONFIDENCE
No cessation lag was used for ozone mortality.	Overestimate	Probably minor. If there is a time lag between changes in ozone exposure and the total realization of changes in health effects then benefits occurring in the future should be discounted. The use of no lag assumes that all mortality benefits are realized in the year of the exposure change and therefore no discounting occurs. This may lead to an overestimate of benefits.	High. Due to the use of short-term studies of ozone mortality, use of a no lag structure is appropriate and supported by the Council HES.
UNCERTAINTIES RELATED TO APPLICATION OF C-R FUNCTIONS			
Application of C-R relationships only to those subpopulations matching the original study population.	Underestimate	Probably minor. The C-R functions for several health endpoints (including PM-related premature mortality) were applied only to subgroups of the U.S. population (e.g. adults 30+) and thus may underestimate the whole population benefits of reductions in pollutant exposures. However, the background incidence rates for these age groups are likely low and therefore would not contribute many additional cases.	High. The baseline mortality and morbidity rates for PM-related health effects are significantly lower in those under the age of 30 (other than neonates).
Application of regionally derived C-R estimates to entire U.S.	Unable to determine based on current information	Probably minor. This is likely to affect morbidity estimates rather than mortality, as mortality	Medium. The differences in the expected changes in health effects

Table 5-11. (Continued)

POTENTIAL SOURCE OF ERROR	DIRECTION OF POTENTIAL BIAS FOR NET BENEFITS ESTIMATE	MAGNITUDE OF IMPACT ON NET BENFITS ESTIMATE	DEGREE OF CONFIDENCE
		estimates are based on studies that include multiple cities. Since morbidity is not as large of a contributor to overall benefits, this is not likely to have a large impact on net benefits.	calculated using different underlying studies can be large. If differences reflect real regional variation, applying individual C-R functions throughout the U.S. could result in considerable uncertainty in health effect estimates.
UNCERTAINTIES RELATED TO HEALTH VALUATION			
Use of a Value-of-a-Statistical-Life (VSL) estimate based on a Weibull distribution of 26 studies	Unable to determine based on current information	Potentially major. Mortality valuation generally dominates monetized benefits.	Medium. The VSL used in this analysis is based on 26 labor market and stated preference studies published between 1974 and 1991. Although there are many more recent studies, including meta-analyses, sensitivity analyses reported above suggest that these alternative sources generate results that are close to the estimates used in the analysis.
Use of cost of illness (COI) estimates to value some morbidity endpoints	Underestimate	Probably minor. Mortality valuation generally dominates monetized benefits; therefore specific estimates used to generate morbidity benefits likely would not have a large impact on net benefits.	Low. Morbidity benefits such as hospital admissions and heart attacks are calculated using COI estimates, which some studies have shown are generally half as

Table 5-11. (Continued)

POTENTIAL SOURCE OF ERROR	DIRECTION OF POTENTIAL BIAS FOR NET BENEFITS ESTIMATE	MAGNITUDE OF IMPACT ON NET BENFITS ESTIMATE	DEGREE OF CONFIDENCE
			much as WTP to avoid the illness. However, WTP estimate are currently not available for all health endpoints.
Benefits transfer for mortality risk valuation, including differences in age, income degree of risk aversion, the nature of the risk, and treatment of latency between mortality risks presented by PM/Cozone and the risks evaluated in the available economic studies.	Unable to determine based on currently available information	Potentially major. The mortality valuation step is clearly a critical element in the net benefits estimate, so any uncertainties can have a large effect.	Medium. Information on the combined effect of these known biases is relatively sparse, and it is therefore difficult to assess the overall effect of multiple biases that work in opposite directions. However, our VSL estimate is based on a distribution of the results of 26 individual studies, which cover a range of characteristics.
Inability to value some quantifiable morbidity endpoints, such as impaired lung function.	Underestimate	Probably minor. Reductions in lung function are a well-established effect, based on clinical evaluations of the impact of air pollutants on human health, and the effect would be pervasive, affecting virtually every exposed individual. However, the lack of a clear symptomatic presentation of the effect, however, could limit individual WTP to avoid lung function decrements.	Low. There currently is no evidence to determine the monetary value of the benefits of avoided lung function reductions.

Table 5-11. (Continued)

POTENTIAL SOURCE OF ERROR	DIRECTION OF POTENTIAL BIAS FOR NET BENEFITS ESTIMATE	MAGNITUDE OF IMPACT ON NET BENFITS ESTIMATE	DEGREE OF CONFIDENCE
UNCERTAINTIES IN FORECASTED DATA SUPPORTING HEALTH EFFECTS ESTIMATES			
Uncertainty in projecting baseline incidence rates	Both	Probably minor. The magnitude varies with the health endpoint. Mortality baseline incidence is at the county level and projected for 5-year increments. Morbidity baseline incidence has varying spatial resolution for year 2000 only.	Medium. The county-level baseline incidence and population estimates were obtained from databases where the relative degree of uncertainty is low. The baseline data for other endpoints are not location specific (e.g., those taken from studies) and therefore may not accurately represent the actual location-specific rates.
Income growth adjustments	Both	Potentially major. Income growth Increases willingness-to-pay Valuation estimates, including mortality, over time.	Medium It is difficult to forecast future income growth, owing to unpredictability of future business and employment cycles. These can have a substantial effect on short term growth rate projections, although over longer periods economic growth rates have tended to converge. The use of data from AEO 2005, however, omits the effect of the most recent economic downturn.
Population projections	Both	Probably minor. The demographics of population forecasting are	Medium. Population projections cannot

Table 5-11. (Continued)

POTENTIAL SOURCE OF ERROR	DIRECTION OF POTENTIAL BIAS FOR NET BENEFITS ESTIMATE	MAGNITUDE OF IMPACT ON NET BENFITS ESTIMATE	DEGREE OF CONFIDENCE
		relatively well-established, however migration estimates are quite uncertain, particularly for specific locations. Overall, we believe that population projections are not likely to vary more than 5 percent at the national level.	adequately account for future population migration due to catastrophic events. Projected population and demographics may not well represent future-year population and demographics.
OTHER UNCERTAINTIES			
Variation in effect estimates reflecting differences in $PM_{2.5}$ composition	Unable to determine based on current information	Unable to determine based on current information	Medium. Epidemiology studies examining regional differences in $PM_{2.5}$-related health effects have found differences in the magnitude of those effects. While these may be the result of factors other than composition (e.g., different degrees of exposure misclassification), composition remains one potential explanatory factor.
Very limited quantification of health effects associated with exposure to air toxics.	Underestimate	Probably minor. Studies have found air toxics cancer risks to be orders of magnitude lower than those of criteria pollutants.	N/A Current data and methods are insufficient to develop (and value) national quantitative estimates of the health effects of these pollutants.
CAAA fugitive dust controls implemented in PM non-attainment	Underestimate	Probably minor. The health and economic benefits of reducing lead	N/A

Table 5-11. (Continued)

POTENTIAL SOURCE OF ERROR	DIRECTION OF POTENTIAL BIAS FOR NET BENEFITS ESTIMATE	MAGNITUDE OF IMPACT ON NET BENFITS ESTIMATE	DEGREE OF CONFIDENCE
areas would reduce lead exposures by reducing the re-entrainment of lead particles emitted prior to 1990. This analysis does not estimate these benefits.		exposure can be substantial (e.g., see section 812 Retrospective Study Report to Congress). However, most additional fugitive dust controls implemented under the *with-CAAA* scenario (e.g., unpaved road dust suppression, agricultural tilling controls, etc.) tend to be applied in relatively low population areas.	

Our primary monetized benefits estimate of avoided premature mortality also assumes a discount rate of five percent. We tested the sensitivity of our primary results by substituting alternative discount rates of three and seven percent.[72] This assumption has a small effect on the benefits estimates; applying a discount rate of seven percent results in benefits that are 6 percent lower than the default and applying a three percent discount rate results in a benefits estimate 6 percent higher than the default.

Qualitative Analysis of Key Factors Contributing to Uncertainty

In addition to the uncertainties outlined above, we identified several other areas of uncertainty related to our health benefits analysis that we did not address quantitatively. This includes sources of uncertainty in our estimation of avoided mortality, not related to across-study variation; application of C-R functions for national benefits estimation; projection of population and baseline incidence rates; and health valuation.

Table 5-11 provides a summary of the key uncertainties related to the Second Prospective health effects modeling analysis. The first column provides a brief description of each key assumption made in the analysis. The second column indicates the direction of the potential bias with respect to the overall net benefits estimate. The third indicates the magnitude of the impact of the potential bias on the net benefits. The Project Team assigns a classification of "potentially major" if a plausible alternative assumption or approach could influence the overall monetary benefit estimate by approximately five percent or more. If an alternative assumption or approach is likely to change the total benefit estimate by less than five percent, the Project Team assigns a classification of "probably minor."73 This assessment is intended to provide readers with a sense for the quantitative impact on the net benefits estimate if an alternate assumption to that selected by the Project Team were to be implemented. Finally, the fourth column provides our level of confidence in the selected assumption, based on our assessment of the available body of evidence. That is, based on the given available evidence, how certain we are that the selected assumption is the most plausible of the alternatives. The Project Team uses the following four qualitative categories to express the degree of confidence in the chosen assumption:

- "High" – the current evidence is plentiful and strongly supports the selected assumption;
- "Medium" – some evidence exists to support the assumption, but data gaps are present; and
- "Low" – there are limited data to support the selected assumption.
- The Project Team uses "N/A" to indicate that the data was so limited that it was excluded from the analysis entirely.

6. Ecological and Other Welfare Benefits

Overview of Approach

Air pollution has important impacts not only on human health, but on a wide range of ecological and environmental resources. Clean Air Act provisions are designed to be protective of human health and the environment, but as a practical matter, because human health impacts are more readily quantified, many of EPA's air pollution analyses have focused much more on human health than on ecological health, aesthetic effects, or natural resource productivity. In general, as science and economics have provided greater insights into the effects of anthropogenic stressors on ecological systems, pursuit of environmental programs targeted on reductions of damage to the environment have become more common. For example, as we noted in the First Prospective, the original motivation for Title IV of the CAAA was addressing the effects of acid rain on ecological resources – it was only after passage that it became clear that these provisions also provide very large human health benefits.

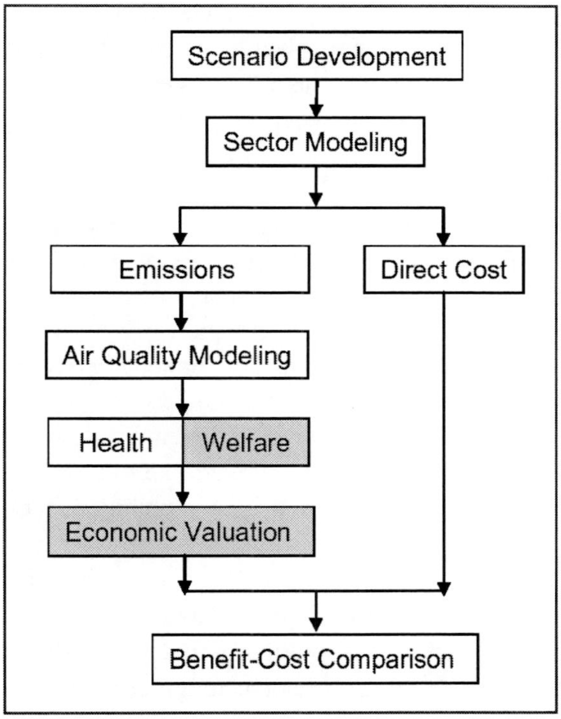

In this chapter, we provide quantitative results for the effects of air pollution on ecological health and natural resources where the science and economic base is strongest, including the lake acidification effects that motivated Title IV, as well as a broad qualitative characterization of effects that are more difficult to quantify. The first portion of this chapter involves taking a broad view of pollutants controlled under the CAAA and their documented effects on ecological systems, both as individual pollutants and, to the extent possible, as one component in multiple-stressor effects on ecosystems and their components. We organize our analysis in terms of major pollutant classes and by the level of biological organization at which impacts are measured (e.g., regional ecosystem, local ecosystem, community, population, organism, etc.). We used a similar strategy in the First Prospective, which has been updated here to reflect new scientific literature published since 1999, but we also supplement the literature review with a new mapping of air pollutant stressors relative to ecological systems that are most sensitive to those stressors – for example, we relate atmospheric deposition of nitrogen to estuarine systems that have been classified as sensitive to marginal nitrogen inputs.

The second portion of the chapter presents the results of a wide range of analyses that quantitatively characterize specific effects of air pollution on ecological systems, as well as other effects on natural and human systems that contribute to economic welfare. We provide quantitative estimates of the benefits of the 1990 CAAA for the following effects:

- Enhanced forest and agricultural plant growth associated with reduced exposure to tropospheric ozone, on a national scale;
- Enhanced visibility in recreational and residential settings associated with reduced particulate matter concentrations, also on a national scale;
- Reduced damage to certain building and structural materials associated with reduced exposure to corrosive air pollutants, such as acid deposition, on a national scale;
- Acidification of freshwater bodies and impairment of timber growth associated with atmospheric nitrogen and sulfur deposition, for a case study area in New York's Adirondack region.

The categories of effects ultimately chosen for quantitative assessment here are necessarily limited by available methods and data. The scope is largely consistent with the recommendations of the Ecological Effects Subcommittee (EES) of the Council, which supported EPA's plans for qualitative characterization of the ecological effects of CAA-related air pollutants, an expanded literature review, national analyses where possible, and a quantitative, ecosystem-level case study of ecological service benefits. As scientific understanding and impact assessment methods grow more comprehensive, however, we expect that the focus of subsequent analyses will continue to broaden, and also yield greater insight on which effects that can be avoided by air pollution controls have the greatest potential ecological and/or economic value.

Because the breadth and complexity of air pollutant-ecosystem interactions do not allow for comprehensive quantitative analysis of all the ecological benefits of the CAAA, we stress the importance of continued consideration of those impacts not valued in this report in policy decision-making and in further technical research. Judging from the geographic breadth and magnitude of the relatively modest subset of impacts that we find sufficiently well-understood to quantify and monetize, it is apparent that the economic benefits of the CAAA's reduction of air pollution impacts on ecosystems are substantial.

Qualitative Characterization of Effects

The First Prospective summarized available information on the ecological effects of criteria pollutants and hazardous air pollutants regulated under the 1990 Clean Air Act Amendments. In this Second Prospective analysis we expand that effort, updating the literature review to reflect published and peer-reviewed research that has become available since the development of the 1999 analysis, through 2008. As data limitations prevent the quantitative assessment of all potential ecological benefits, the goal of this effort is to provide a broad characterization of the range of effects of major air pollutants on ecological endpoints.

Table 6-1. Classes of Pollutants and Ecological Effects.

POLLUTANT CLASS	MAJOR POLLUTANTS AND PRECURSORS	ACUTE EFFECTS	LONG-TERM EFFECTS
Acidic deposition	Sulfuric acid, nitric acid *Precursors:* Sulfur dioxide, nitrogen oxides	Direct toxic effects to plant leaves and aquatic organisms.	Progressive deterioration of soil quality due to nutrient leaching. Forest health decline. Acidification of surface waters. Reduction in acid neutralizing capacity in lakes and streams. Enhancement of bioavailability of toxic metals (aluminum) to aquatic biota.
Nitrogen Deposition	Nitrogen compounds (*e.g.*, nitrogen oxides)		Nitrogen saturation of terrestrial ecosystems, causing nutrient imbalances and reduced forest health. Soil and water acidification. Reduction in acid neutrallizing capacity in lakes and streams. Progressive nitrogen enrichment of coastal estuaries causing eutrophication. Changes in the global nitrogen cycle.
Ozone	Tropospheric ozone *Precursors:* Nitrogen oxides and volatile organic compounds (VOCs)	Direct toxic effects to plants.	Alterations of ecosystem wide patterns of energy flow and nutrient cycling; community changes.
Hazardous Air Pollutants (HAPs)	Mercury, dioxins	Direct toxic effects to animals.	Conservation of mercury and dioxins in biogeochemical cycles and accumulation in the food chain. Sublethal impacts.

Ecosystem impacts can be organized by the pollutants of concern and by the level of biological organization at which impacts are directly measured. We address both dimensions of categorization in this overview. Table 6-1 summarizes the major pollutants of concern, and the documented acute and long-term ecological impacts associated with them.

The following discussion provides more specific information on ecological effects of each pollutant class, including information on sources, sensitive ecosystems, and summary tables of effects organized by level of biological organization.

Acidic Deposition

The predominant chemicals associated with acidic precipitation are sulfuric and nitric acid (H_2SO_4 and HNO_3). These strong mineral acids are formed from sulfur dioxide (SO_2) and nitrogen oxides (NO_x) in the atmosphere. Sulfur compounds are emitted from anthropogenic sources in the form of SO_2 and, to a lesser extent, primary sulfates, principally from coal and residual-oil combustion and a few industrial processes. The principal anthropogenic source of NO_x emissions is fuel combustion. In the atmosphere, SO_2 and NO_x are converted to sulfates and nitrates, transported over long distances, and deposited over large areas downwind of urban areas or point sources.

Acidification of ecosystems has been shown to cause direct toxic effects on sensitive organisms as well as long-term changes in ecosystem structure and function. The effects of acidification can be seen at all levels of biological organization in both terrestrial and aquatic ecosystems. Adverse effects in terrestrial ecosystems include acutely toxic impacts of acids on terrestrial plants and, more commonly, chronic acidification of terrestrial ecosystems leading to nutrient deficiencies in soils, aluminum mobilization, and decreased health and biological productivity of forests. These effects can lead to changes in individual plant survival, as well as changes in forest populations and communities.

In aquatic ecosystems, acidification-induced effects are mediated by changes in water chemistry including reductions in Acid Neutralizing Capacity[74] (ANC) and increased availability of aluminum (Al^{3+}), which in turn can cause increased mortality in sensitive species, changes in community composition, and changes in nutrient cycling and energy flows. Acidic deposition has resulted in increased acidity in surface waters, especially in areas where acid buffering capacity of soils is reduced and nitrate and sulfate have leached from upland areas. While many fish species are acid-sensitive, the main lethal agent is the increase in dissolved aluminum that occurs with falling pH levels.

Acid-sensitive ecosystems include those with high acidic deposition and low acid neutralizing capacity. Many of these ecosystems occur downwind of emission sources, often in mountainous areas where soils are thin and poorly buffered. High elevation sites are also more vulnerable because mountain fog is often more acidic than rain.

Table 6-2 provides a summary of the potential ecological effects of acidification.

Nitrogen Deposition

Along with its role in acidification of ecosystems, nitrogen deposition also affects nitrogen biogeochemistry, which in turn affects the health of forest and coastal ecosystems. Nitrogen is a naturally occurring element, and is essential to both plant and animal life, but combustion processes cause this nitrogen to be "fixed" – that is, converted from the unreactive N2 form to a reactive form such as nitrate (NO3) or ammonia (NH_3). The availability of reactive nitrogen limits plant growth in many terrestrial ecosystems and is generally the limiting nutrient in marine and coastal waters as well.

By 1990, human activities had more than doubled the amount of reactive nitrogen available annually to living organisms. At present, more than 50 percent of the annual global reactive nitrogen emissions are generated directly or indirectly by human activities. Ammonia emissions to the atmosphere occur largely via volatilization from animal wastes. Anthropogenic nitrogen oxide (NO_x) emissions to the atmosphere are generally a result of fossil fuel combustion, with electric power generation and automobiles as the largest two sources.

Because most terrestrial and coastal ecosystems are nitrogen limited, increased supply of nitrogen in terrestrial systems can stimulate uptake by plants and microorganisms, and increase biological productivity. Moderate levels of nitrogen input can have a "fertilizing" effect, similar to the application of nitrogen fertilizer frequently used in timber production or agriculture. In the long run, however, chronic nitrogen deposition adversely affects organisms, communities, and biogeochemical cycles of watersheds and coastal waters. Biogeochemical cycles change when the nutrient balance is disrupted by excess nitrogen because nitrogen is an important nutrient in biological systems.

Table 6-2. Effects of Acidification on Natural Systems at Various Levels of Organization.

SPATIAL SCALE	TYPE OF INTERACTION	EXAMPLES OF EFFECTS	
		FOREST ECOSYSTEMS	STREAMS AND LAKES
Molecular and cellular	Chemical and biochemical processes	Damages to epidermal layers and cells of plants through deposition of acids; alteration of stomatal activity.	Decreases in pH and increases in aluminum ions cause pathological changes in structure of gill tissue in fish.
Organism	Direct physiological response	In trees, increased loss of nutrients via foliar leaching.	Hydrogen and aluminum ions in the water column impair regulation of body ions.
	Indirect effects: Acidification can indirectly affect response to altered environmental factors or alterations of the organism's ability to cope with other kinds of stress.	Cation depletion in the soil causes nutrient deficiencies in plants. Concentrations of aluminum ions in soils can reach phytotoxic levels. Increased sensitivity to other stress factors including pathogens and frost. In birds, possible calcium limitation and growth reduction.	Aluminum ions in the water column can be toxic to many aquatic organisms through impairment of gill regulation.
Population	Change of population characteristics like productivity or mortality rates.	Decrease of biological productivity of sensitive organisms. Selection for less sensitive organisms. Microevolution of resistance.	Decrease of biological productivity and increased mortality of sensitive organisms. Selection for less sensitive organisms. Microevolution of resistance.
Community	Changes of community structure and competitive patterns.	Alteration of competitive patterns. Selective advantage for acid-resistant species. Loss of acid sensitive species and organisms. Decrease	Alteration of competitive patterns. Selective advantage for acid-resistant species. Loss of acid sensitive

Table 6. (Continued)

SPATIAL SCALE	TYPE OF INTERACTION	EXAMPLES OF EFFECTS	
		FOREST ECOSYSTEMS	STREAMS AND LAKES
		in productivity. Decrease of species richness and diversity. Decline in Sugar Maple and red spruce in Eastern U.S. and Canadian forests.	species and organisms. Decrease in productivity. Decrease in species richness and diversity.
Local Ecosystem (e.g., landscape element)	Changes in nutrient cycle, hydrological cycle, and energy flow of lakes, wetlands, forests, grasslands, etc.	Progressive depletion of nutrient cations in the soil. Increase in the concentration of mobile aluminum ions in the soil.	Acidification of lakes and streams. Decrease in acid neutralizing capacity.Persistent acidic conditions in lakes and streams in some regions, despite reduction in sulfate deposition.
Regional Ecosystem (e.g., watershed)	Biogeochemical cycles within a watershed. Region-wide alterations of biodiversity.	Leaching of sulfate, nitrate, aluminum, and calcium to streams and lakes. Change in sulfur and nitrogen biogeochemistry in northeastern forests.	Regional acidification of aquatic systems due to high deposition rates and nitrogen saturation of terrestrial ecosystems and increased nitrate leaching to surface waters. Persistent acidic conditions in lakes and streams in some regions, despite reduction in sulfate deposition.

Because fresh waters are generally not nitrogen limited, the addition of nitrogen does not lead to excessive eutrophication as it does in coastal waters. Coastal waters are an extraordinarily important natural resource, providing spawning grounds/nurseries for fish and shellfish, foraging and breeding habitat for birds, and generally contributing greatly to the productivity of the marine environment. Critical to the health of coastal waters is an appropriate balance of nutrients. If present in mild or moderate quantities, nitrogen enrichment of coastal waters can cause moderate increases in productivity, leading to neutral or positive changes in the ecosystem. However, because coastal waters are generally nitrogen limited, too much nitrogen leads to excess production of algae, decreasing water clarity and reducing concentrations of dissolved oxygen, a situation referred to as eutrophication.

Table 6-3 summarizes the potential effects of nitrogen deposition on ecosystem structure and function.

Table 6-3. Effects of Nitrogen Deposition on Natural Systems at Various Levels of Organization.

SPATIAL SCALE	TYPE OF INTERACTION	EXAMPLES OF EFFECTS	
		FOREST ECOSYSTEMS	ESTUARINE ECOSYSTEMS
Molecular and cellular	Chemical and biochemical processes.	Increased uptake of nitrogen by plants and microorganisms. With chronic exposure, reduced stomatal activity and photosynthesis in some species.	Increased assimilation of nitrogen by marine plants, macroalgae, and microorganisms.
Organism	Direct physiological response.	Increases in leaf size of terrestrial plants. Increase in foliar nitrogen concentration in major canopy trees. Change in carbon allocation to various plant tissues.	Increase in algal growth.
	Indirect effects: Response to altered environmental factors or alterations of the organism's ability to cope with other kinds of stress.	Decreased resistance to biotic and abiotic stress factors including pathogens, insects, and frost. Disruption of plant-symbiont relationships with mycorrhizal fungi.	Injuries to marine fauna through depletion of oxygen in the water column. Loss of physical habitat due to increased macroalgal biomass and loss of seagrass beds. Injury and habitat loss through increased shading by macroalgae.
Population	Change of population characteristics like productivity or mortality rates.	Increase in biological productivity and growth rates of some species. Increase in pathogens.	Increase in algal and macroalgal biomass.
Community	Changes of community structure and competitive patterns.	Alteration of competitive patterns. Selective advantage for fast growing species and organisms that efficiently use additional nitrogen. Loss of species adapted to nitrogen-poor or acidic environments. Increase in weedy species or parasites.	Excessive algal growth. Changes in species composition with increase in algal and macroalgal species and decrease or loss of seagrass beds. Loss of species sensitive to low oxygen conditions.
Local Ecosystem (e.g.,	Changes in nutrient cycle, hydrological cycle, and energy flow of lakes,	Changes in the nitrogen cycle. Progressive nitrogen saturation. Mobilization of nitrate	Changes in the nitrogen cycle. Increased algal growth leading to depletion of oxygen,

Table 6-3. (Continued)

SPATIAL SCALE	TYPE OF INTERACTION	EXAMPLES OF EFFECTS	
		FOREST ECOSYSTEMS	ESTUARINE ECOSYSTEMS
landscape element)	wetlands, forests, grasslands, etc.	and aluminum in soils. Loss ofcalcium and magnesium from soil. Change in organic matter decomposition rate.	increased shading of seagrasses. Reduced water clarity and dissolved oxygen levels.
Regional Ecosystem (e.g., watershed)	Changes in biogeochemical cycles within a watershed. Region-wide alterations of biodiversity.	Leaching of nitrate and aluminum from terrestrial sites to streams and lakes. Acidification of soils and waterbodies. Increased emission of greenhouse gases from soils to atmosphere. Change in nutrient turnover and soil formation rates.	Additional input of nitrogen from nitrogen-saturated terrestrial sites within the watershed. Regional decline in water quality in waterbodies draining large watersheds (e.g. Chesapeake Bay). Changes in the regional-scale nitrogen cycle.
Global Ecological System	Changes in global biogeochemical cycles; increased availability of reactive nitrogen to plants.	Increased input of reactive nitrogen; loss of soil nutrients. Nitrogen saturation and leaching throughout forests in northeastern United States and Western Europe. Acidification of surface waters.	Greatly increased transfer of nitrogen to coastal ecosystems; change in structure and function of estuarine and nearshore systems.

Tropospheric Ozone

Ozone is a secondary pollutant formed through the oxidation of volatile organic compounds (VOCs) in the presence of oxides of nitrogen. Ozone is one of the most powerful oxidants known but its impacts have been little studied in faunal species. The limited available research has shown a variety of pulmonary impacts to specific mammalian and avian species. In contrast, ozone's impacts on plants are much better understood. Documented effects on forest trees include visible foliar damage, decreased chlorophyll content, accelerated leaf senescence, decreased photosynthesis, increased respiration, altered carbon allocation, water balance changes, and damage to epicuticular wax. These can lead to changes in canopy structure, carbon allocation, productivity, and fitness of trees.

Ozone sensitivity of plants varies between species, with evergreen species tending to be less sensitive to ozone than deciduous species, and with most individual deciduous trees being less sensitive than most annual plants. However, there are exceptions to this broad ranking scheme, and there can be variability not only between species but even between clones of some trees and within cultivars. Life stage also matters: in general, mature deciduous trees tend to be more sensitive than seedlings, while the reverse is more typical for evergreen trees.

Impacts to plant communities may occur as a result of ozone exposure, although such effects have not been studied as extensively due to ecosystem complexity and the long timeframes involved. Experiments with an early successional plant community found that ozone reduced vegetative cover, vertical density, species richness, and evenness relative to the control, although differences were less pronounced in a drought year. Other observed community level effects include reduced competitive ability of sensitive species, changed soil microbial communities, and altered species composition and relative abundance.

Table 6-4 summarizes the potential effects of ozone exposure on ecosystems.

Table 6-4. Effects of Ozone on Natural Systems at Various Levels of Organization.

SPATIAL SCALE	TYPE OF INTERACTION	EXAMPLES OF EFFECTS
Molecular and cellular	Chemical and biochemical processes.	Oxidation of enzymes of plants, generation of toxic reactive oxygen species (hydroxyl radicals). Disruption of the membrane potential. Reduced photosynthesis and nitrogen fixation. Increased apoptosis.
Organism	Direct physiological response.	Visible foliar damage, premature needle senescence, altered carbon allocation, and reduced growth rates.
	Indirect effects: Response to altered environmental factors or alterations of the organism's ability to cope with other kinds of stress.	Increased sensitivity to biotic and abiotic stress factors such as pathogens and frost. Disruption of plant-symbiont relationship (mychorrhizae), and symbionts.
Population	Change of population characteristics like productivity or mortality rates.	Reduced biological productivity and reproductive success. Selection for less sensitive organisms. Potential for microevolution for ozone resistance.
Community	Changes of community structure and competitive patterns.	Alteration of competitive patterns. Loss of ozone sensitive species and organisms leading to reduced species richness and evenness. Reduction in productivity. Changes in microbial species composition in soils.
Local Ecosystem (e.g., landscape element)	Changes in nutrient cycle, hydrological cycle, and energy flow of lakes, wetlands, forests, grasslands, etc.	Alteration of ecosystem-wide patterns of energy flow and nutrient cycling (e.g., via alterations in litter quantity, litter nutrient content, and degradation rates; also via changing carbon fluxes to soils and carbon sequestration in soils).
Regional Ecosystem (e.g., watershed)	Biogeochemical cycles within a watershed. Region-wide alterations of biodiversity.	Potential for region-wide phytotoxicological impacts and reductions in net primary production.

Hazardous Air Pollutants

Hazardous air pollutants (HAPs) are a general category of toxic substances covered under Title III of the Clean Air Act, which lists 189 HAPs. Of these 189 substances, the best understood in terms of the potential for adverse ecological impacts include mercury, polychlorinated biphenyls (PCBs), dioxins, and dichlorodiphenyl-trichloroethane (DDT). The use of PCBs and DDT was effectively illegal in the United States prior to 1990 (EPA 1992), and there are currently no plans for additional CAAA regulations of these compounds (Federal Register Unified Agenda 1998). With respect to mercury and dioxins, regulatory actions have reduced, but have not eliminated, anthropogenic emissions. This section discusses environmental effects associated with these two HAPs.

Mercury

Mercury (Hg) is a toxic element found ubiquitously throughout the environment. About 50-80 percent of total emissions originate from anthropogenic sources, including fossil fuel combustion, leaks from industrial activities, and the disposal or incineration of wastes.

Mercury is generally released in its elemental and inorganic forms. However, it can undergo various transformations in the environment, and its chemical form determines not only its environmental fate but also its potency as a toxicant. From a biological perspective, the most hazardous form of mercury is methylmercury both because of its bioaccumulation and biomagnification potential, and also because organic forms of mercury (including methylmercury) are the most toxic. Adverse effects on wildlife include neurotoxicity as well as reproductive, behavioral, and developmental effects. These types of effects have been observed in laboratory studies of mammals, birds, fish, and aquatic invertebrates. While species sensitivity varies, within a species the early life stages are generally the most sensitive.

Dioxins

Polychlorinated dibenzo-p-dioxins (PCDDs) are a group of 75 organochlorine compounds, often referred to as dioxins. Although dioxins can be produced through natural events such as forest fires and volcanic eruptions, most environmental inputs are anthropogenic in origin. EPA categorizes dioxin sources into five broad groups: combustion; metals smelting, refining, and processing sources; chemical manufacturing; biological and photochemical processes; and reservoir sources (for example urban runoff).

Dioxins and related compounds are thought to exert most of their toxic effects through interaction with the aryl hydrocarbon receptor (AhR). In laboratory studies, particularly of rodents, some dioxins have been shown to cause reproductive toxicity, neurotoxicity, immune suppression, increased inflammatory responses, and cancer. Fish are among the most sensitive species to the effects of dioxin, and early life stages are the most vulnerable. The risk that dioxins pose to other wildlife is difficult to assess because both laboratory and field studies are few.

Dioxins are extremely stable chemicals with a persistence that is measured in decades. Dioxins are subject to photochemical degradation, but since the penetration of light into soils and many natural water bodies is limited, this degradation is slow. Because of dioxins' toxicity and persistence, their presence is likely to be an issue of concern for decades.

Distribution of Air Pollutants in Sensitive Ecosystems of the United States

This section describes the spatial and temporal trends of air pollutants regulated by the CAAA, highlighting their distribution against sensitive ecosystems across the United States. This information provides useful context regarding the geographic distribution of potential ecological benefits of the CAAA, particularly for the ecological endpoints described above for which data are not available to quantify impacts.

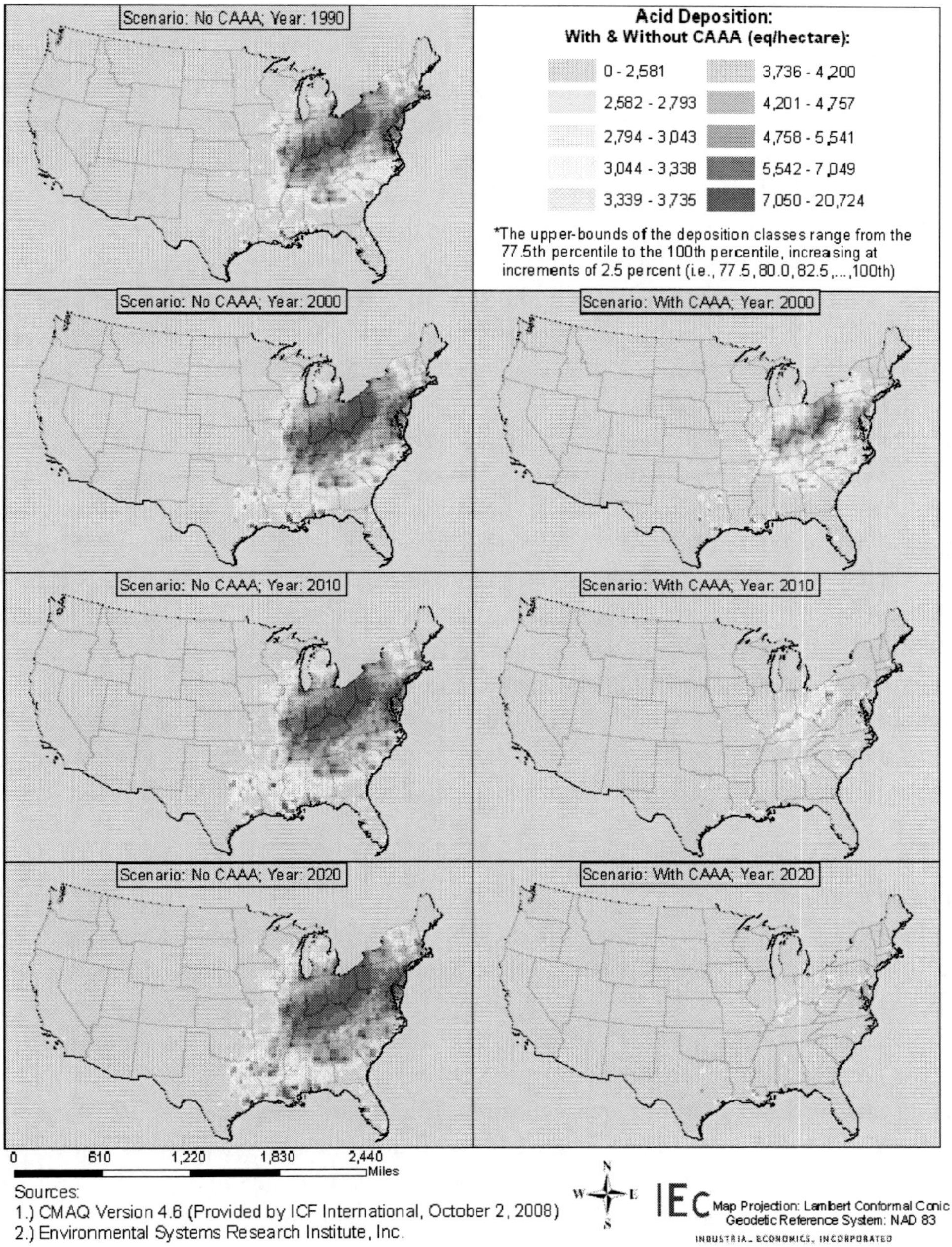

Figure 6-1. Combined Nox and Sox Deposition Estimates for 1990, 2000, 2010, and 2020 with and without the CAAA.

The maps presented illustrate changes in forecast pollutant levels under the current, baseline scenario (with the CAAA) as compared to the counterfactual scenario (without the CAAA). The three pollutant classes considered are: acidic deposition, nitrogen deposition, and tropospheric ozone. Data are not available to map the distribution of HAPs. The pollutant exposure maps presented in this discussion were created using data from the Community Multiscale Air Quality Modeling System (CMAQ) Version 4.6, which estimates tropospheric ozone concentrations as well as deposition in kilograms per hectare for acidic deposition and total nitrogen.[75]

Acidic Deposition

As described in the previous section, ecosystem sensitivity to acid deposition occurs in areas with low ANC. High elevation sites tend to be more vulnerable because of thin, poorly buffered soils coinciding with acidic deposition from rain, snow, and fog. Acid- sensitive areas in the U.S. include the southern Blue Ridge Mountains of eastern Tennessee, western North Carolina and northern Georgia; the mid Appalachian Region of eastern West Virginia, western Virginia and central Pennsylvania; New York's Catskill and Adirondack Mountains; the Green Mountains of Vermont; the White Mountains of New Hampshire, and areas of the Upper Midwest (Wisconsin and Michigan).[76] Montane areas in the Adirondacks, Northern New England, and the Appalachian region have experienced acidification of surface waters and soils, as well as forest decline.

Figure 6-1 presents acidic deposition from 1990 through 2020 for both *with-* and *without-CAAA* scenarios. Acid deposition estimates are expressed as equivalents per hectare (eq/ha).[77] Under both regulatory scenarios, acidic deposition is highest in western Pennsylvania, southern Ohio and Indiana, western West Virginia, and northern Kentucky. Without the CAAA, acidic deposition in these areas increases over time. Further, acidic deposition increases over time in the areas surrounding these hotpots. By 2020, significant portions of the Northeast, Midwest, and South are projected to have elevated levels of acidic deposition. Hotspots also exist in eastern Texas and southern Louisiana.

As shown in the right column of Figure 6-1, with the CAAA acidic deposition levels lessen in and around the areas with the highest acidic deposition. By 2020, elevated acidic deposition levels are primarily limited to much smaller areas in the Midwest, Northeast, and Gulf Coast.

Nitrogen Deposition

Atmospheric nitrogen deposition is highest in the northeastern and eastern central regions of the U.S. Elevated nitrogen deposition in the western and southern United States is limited to areas in the vicinity of large nitrogen sources (e.g., livestock production areas), high-elevation areas on which cloud droplet deposition may contribute substantial nitrogen inputs, and urban areas with relatively high levels of NOx emissions.

Figure 6-2 presents total nitrogen deposition from years 1990 through 2020 for both the *with-CAAA* and *without-CAAA* scenarios. In general, total nitrogen deposition is less than 24 kg/hectare in the conterminous U.S. for each year and regulatory scenario presented. However, "hot spots" exist across the U.S. where meteorological conditions and/or high nitrogen emissions contribute to relatively high deposition rates. Two particularly significant hot spots for nitrogen deposition are located in southern Louisiana and eastern North Carolina. Total nitrogen deposition is estimated to increase in both hot spots over time regardless of the regulatory scenario. Outside of the two hot spots, total nitrogen deposition is highest without the CAAA in the Ohio River Valley (i.e., western Pennsylvania, southern Ohio and Indiana, western West Virginia, and northern Kentucky). Over time, the total nitrogen deposition increases around the Ohio River Valley without the CAAA and decreases slightly with the CAAA. Outside of the Ohio River Valley, nitrogen deposition with the CAAA decreases slightly over time in the eastern U.S. In the western U.S., total nitrogen deposition with the CAAA remains relatively constant over time.

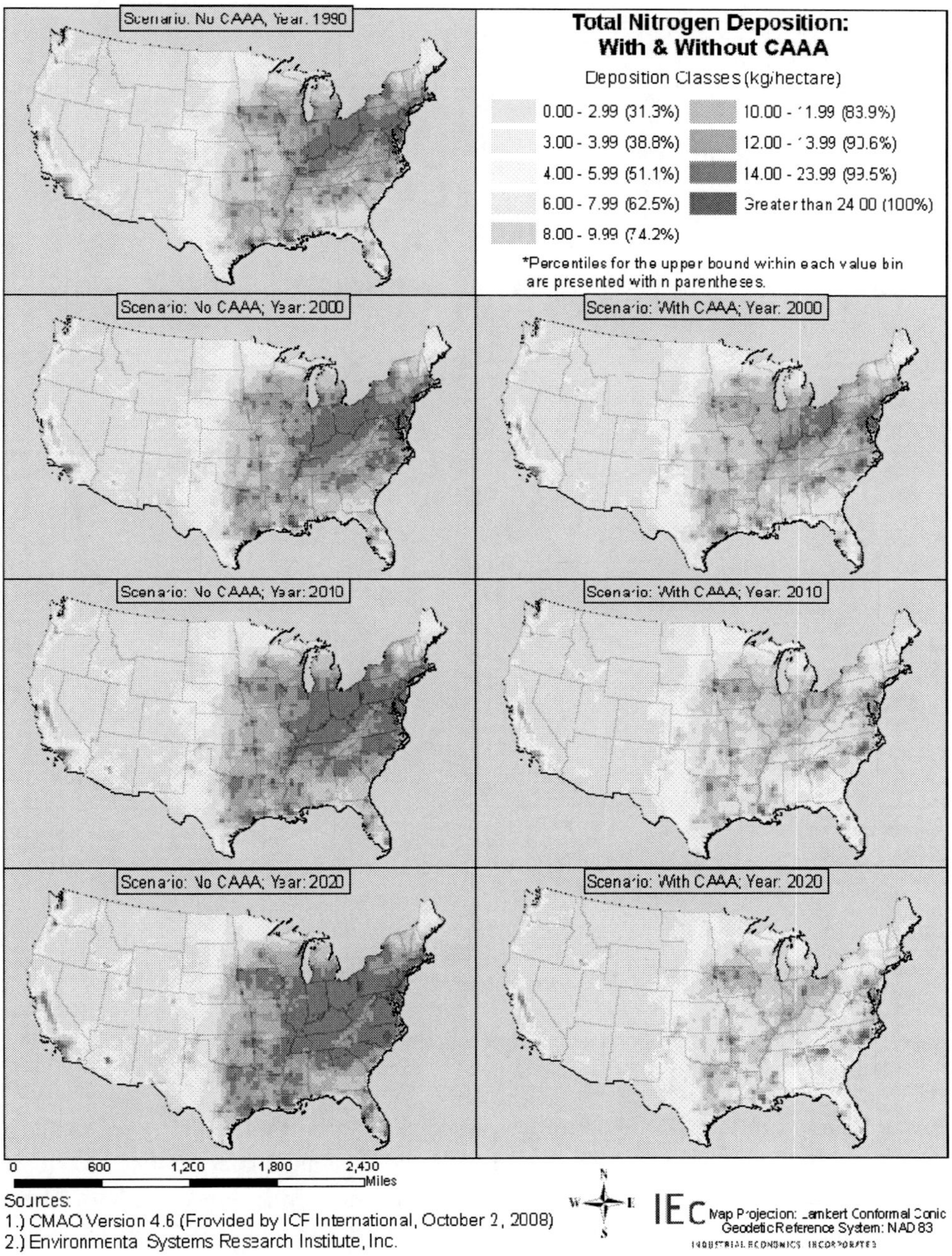

Figure 6-2. Total Nitrogen Deposition Estimates for 1990, 2000, 2010, and 2020 with and without the CAAA[78,79].

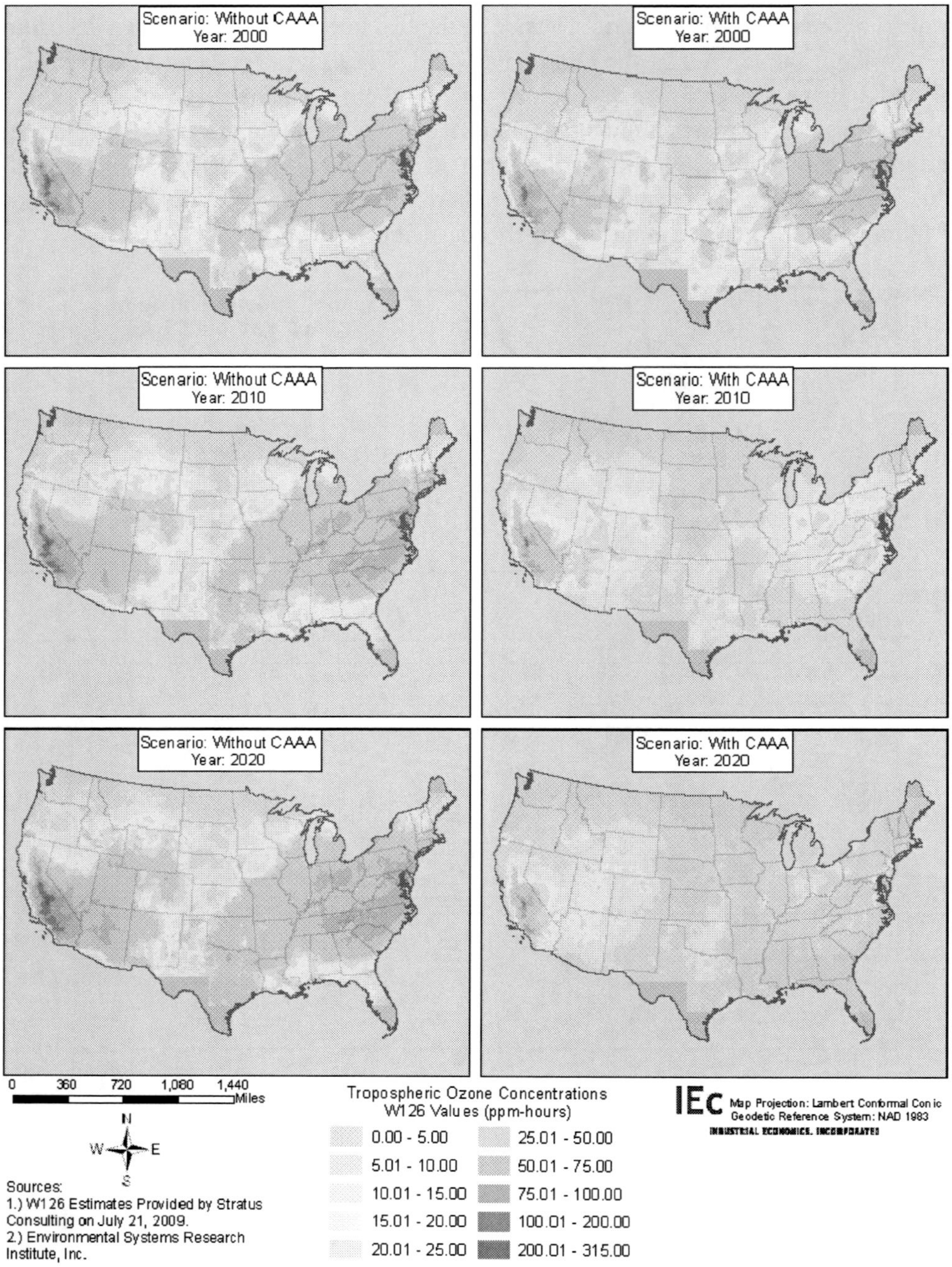

Figure 6-3. W126 Cumulative Tropospheric Ozone Season Measures for 2000, 2010, and 2020 with and without the CAAA.

Estuarine areas in the Northeast are less susceptible to injury from nitrogen loading than estuaries in other parts of the country due to the rapid flushing characteristics of estuaries in this region. Estuaries along the Southeastern Coast, Gulf Coast, and Southern California Coast experience the greatest reduction in total nitrogen deposition. Total nitrogen deposition along the West Coast, with the exception of southern California, is relatively low in the absence of the CAAA.

Tropospheric Ozone

Areas within the U.S. with elevated tropospheric ozone levels include the Northeast, mid-Atlantic, Midwest, and California. Combined ozone concentrations are reported for the May through September period as ozone levels tend to increase during the spring and summer. Figure 6-3 presents combined cumulative ozone season (W126) values for the May through September period for both the *with-CAAA* and *without-CAAA* scenarios. The W126 metric is a weighted sum of hourly concentrations observed between 8 a.m. and 8 p.m. where hourly weights are a function of the hourly ozone concentration observed.

In general, tropospheric ozone concentrations increase over time without the CAAA and decrease over time with the CAAA. Elevated ozone concentrations are present in California, mid-Atlantic states, and Corn Belt states in 2000 both with and without the CAAA; ozone concentrations are, however, slightly less with the CAAA in 2000. In 2000, ozone hot spots are present in southern California, central Ohio, portions of Virginia, North Carolina, and South Carolina, and western Tennessee. Without the CAAA, these hot spots grow in size and magnitude. Under the *with-CAAA* scenario, the hot spots decrease in size and magnitude. By 2020, the combined W126 values for nearly the entire conterminous U.S. (outside of California) are less than 15 ppm-hours. Tropospheric ozone concentrations within the California hot spot are reduced to 25 to 75 ppm-hours.[80]

As noted in the previous section, elevated tropospheric ozone levels may negatively affect plants in a number of ways, including reducing plant photosynthesis and increasing leaf senescence leading to reduced plant growth and productivity. Given the potential effects of elevated tropospheric ozone concentrations on plant growth, forested and cropland areas across the U.S. are considered particularly sensitive to the effects of elevated tropospheric ozone. It follows that these same areas also stand to benefit the most from reduced tropospheric ozone concentrations due to the implementation of the CAAA. In particular, forested ecosystems in the San Bernardino and Sierra Nevada Mountains of California have suffered ecological damages attributed to elevated ozone levels. Forests in the southern portions of the Midwest and Northeast regions and the Southeast region (except the southernmost areas where ozone concentrations are relatively low without the CAAA) are also expected to benefit from reductions in tropospheric ozone due to the implementation of the CAAA. In addition, crops in California are expected to benefit the most from the implementation of the CAAA. The cropland areas in California are located almost entirely within the tropospheric ozone hot spot. Other cropland areas expected to benefit from reduced tropospheric ozone concentrations associated with the implementation of the CAAA include the Corn Belt region, the southern portion of the Midwest region, the Mississippi Valley, Texas, and Oklahoma.

Quantified Results: National Estimates

Agriculture and Forest Productivity Effects

A significant body of literature exists addressing the effects of tropospheric ozone on plants, including commercial tree species and agricultural crops, as noted in the previous section. In general, elevated levels of tropospheric ozone have been shown to reduce overall plant health and growth by reducing photosynthesis and altering carbon allocation. Methods and data also exist to estimate the magnitude of plant growth reductions due to elevated tropospheric ozone levels, based on laboratory studies that developed exposure-response functions describing the functional relationship between plant yield and ozone exposure for a variety of plant species.[81] Applying exposure- response functions, this analysis estimates yield losses in agricultural crops and commercial tree species under the counterfactual, *without-CAAA* scenario relative to the baseline, *with-CAAA* scenario. Relative yield losses (i.e., reductions in crop and tree yield under the counterfactual scenario relative to the baseline scenario) measure the amount crop and tree yields would be reduced in the absence of CAAA regulations, and therefore, indicate a benefit of the CAAA.[82]

Table 6-5 provides a summary of estimated relative yield losses by crop/forest type and year. Relative yield losses indicate a benefit of the CAAA; the larger the relative yield loss without the CAAA, the greater the crop or tree yield with the CAAA.

Table 6-5. Minimum, Maximum, and Average Annual Relative Yield Losses across all Fasom Subregions for Crops and all Fasom Regions for Trees by Year (2000, 2010, 2020).

CROP/FOREST TYPE	2000			2010			2020		
	MINIMUM	MAXIMUM	AVERAGE	MINIMUM	MAXIMUM	AVERAGE	MINIMUM	MAXIMUM	AVERAGE
Barley	0.00%	0.02%	0.01%	0.00%	0.06%	0.02%	0.00%	0.07%	0.02%
Corn	0.00%	1.12%	0.18%	0.00%	3.07%	0.44%	0.00%	3.45%	0.56%
Cotton	0.00%	6.60%	1.15%	0.00%	16.67%	3.00%	0.00%	20.31%	3.81%
Oranges	0.00%	1.95%	0.09%	0.00%	4.68%	0.25%	0.00%	7.87%	0.43%
Potato	0.00%	6.17%	1.76%	0.00%	17.54%	4.99%	0.00%	20.80%	6.50%
Rice	-0.08%	0.14%	0.00%	0.00%	1.03%	0.11%	0.00%	1.66%	0.18%
Sorghum	0.00%	0.87%	0.14%	0.00%	2.17%	0.35%	0.00%	2.65%	0.47%
Soybean	0.00%	3.60%	1.24%	-0.55%	11.73%	3.07%	0.00%	12.74%	4.26%
Processing Tomatoes	0.00%	1.82%	0.31%	0.00%	5.54%	0.96%	0.00%	8.21%	1.47%
Spring Wheat	0.00%	1.50%	0.06%	0.00%	3.67%	0.15%	0.00%	6.98%	0.28%
Winter Wheat	0.00%	6.53%	1.00%	0.00%	18.23%	2.49%	0.00%	19.23%	3.29%
Hardwood Forests	1.60%	7.16%	5.06%	4.20%	19.12%	13.86%	6.61%	23.04%	16.68%
Softwood Forests	0.06%	3.85%	1.77%	0.25%	10.49%	4.88%	0.42%	12.27%	6.11%

Note: Negative relative yield losses indicate yield reductions with the CAAA. For example, the minimum estimate for soybeans in 2010 reflects an estimated relative yield loss of -0.55 percent. The negative relative yield loss is due to reductions in W126 ozone metric values under the counterfactual, no CAAA scenario in Florida in September of 2010 (the growing period for soybeans in Florida is roughly mid-July through September). In other words, ozone exposure is greater under the with-CAAA scenario for that month and region and, therefore, a net increase in soybean yield occurs assuming a rollback of the CAAA. Ozone concentrations are lower under the baseline, with CAAA scenario in Florida for all other months in 2010.

In addition, Figures 6-4 and 6-5 provides maps of the crop-subregion-specific and tree-region-specific relative yield losses for two representative species: potatoes and softwood trees. The results presented generally follow the temporal and spatial pattern of ozone concentration reductions attributable to the CAAA, as outlined in Chapter 4, with reductions in tropospheric ozone concentrations being greatest along the East Coast, particularly the

Southeast, in the Midwest (within the Ohio River Valley), and in California. Several other factors also affect yield changes in crops and trees, including sensitivity to ozone, geographic distribution, growing period length, and the specific time of year the growing period occurs. Potatoes and softwoods, as indicated in Table 6-5, suffer relatively larger changes in growth than some other species in our analysis, and yield losses tend to increase over time as differences in ozone concentrations increase between the *withCAAA* and *without-CAAA* scenarios. Across all crops, the largest relative yield losses for both crops and trees occur in the Southeast, frequently in Virginia, North Carolina, South Carolina, and Tennessee.

Commercial timber and agriculture operations generally manage their land to maximize profits. As such, changes in crop yields between the baseline and counterfactual scenarios may affect the distribution of commercial species planted; for example, landowners may shift production towards plants that are less sensitive to elevated ozone concentrations under the counterfactual scenario. This may occur at the individual plant level, replacing one crop or tree species for another with a higher growth rate; or, it may occur at the community level, converting agricultural lands to timberlands, or vice versa, to adjust for combined yield losses to agricultural crops and commercial tree species.

Changes in the distribution and yield of crop and tree species may in turn affect the supply of and demand for agricultural crops and commercial tree species, resulting in changes in the welfare of consumers and within agricultural and timber sectors of the economy. To quantify this economic benefit of cleaner air, we used the Forest and Agriculture Sector Optimization Model (FASOM). FASOM development was funded by EPA's Climate Economics Branch (CEB) and other EPA, U.S. government, and nongovernmental funders over several decades as a partial equilibrium tool to evaluate the welfare and market impacts of public policies affecting agriculture and forestry. The model simulates biophysical and economic processes affecting land management and land allocation decisions over time to potentially competing agriculture and forest activities. Although the latest version of FASOM was developed to evaluate climate and biofuels policies, the model is capable of assessing a broad range of factors that might affect plant growth; for this project, we worked with the model's developers to develop input files to characterize the impact of ozone on plant and tree growth at a regional and crop-specific level, using the exposure-response results described above.[83]

Although FASOM has been widely applied to agricultural sector analysis and has been peer reviewed in many contexts, it has not to date been subject to a validation exercise comparing the model results for an historical period to historical data for that period.[84] As a result, the performance of the model in forecasting future agricultural sector effects, such as those estimated for this study, has not yet been assessed. Two other potential limitations may pertain in EPA's application of FASOM for this study. First, FASOM adopts a model simulation approach which assumes perfect foresight by economic actors in the agricultural sector. A perfect foresight assumption may be of concern for some long-term analyses, but is likely to be less problematic for this study because our time horizon extends only to 2020. Furthermore, USDA projections of commodity prices and outputs also extend nearly to 2020, and FASOM's projections for their base case agree well with the USDA projections. As a result, the effect of perfect foresight on model outcomes in the present study is reduced.[85] A second potential limitation of FASOM is its approach to estimating the sensitivity of imports to changes in domestic prices. Although FASOM is not a full international model, it does incorporate an import elasticity estimate for the largest and most important commodity crops.

This allows the model to capture, for example, increases in agricultural imports to the US under a scenario in which domestic crop prices are projected to rise. For a number of minor crops, traded in very small quantities, however, FASOM holds imports fixed. The effect of this factor on our results is not clear, but we estimate that a more flexible import sector for these much less important crops would have only a minor effect on our estimates of the net benefits of reducing ozone exposure for US crops. We expect the directional bias of holding minor crop imports fixed, while small, would be to slightly reduce our estimates of the net welfare benefit of reducing ozone exposure, and thereby improving productivity, of domestic agricultural crops.

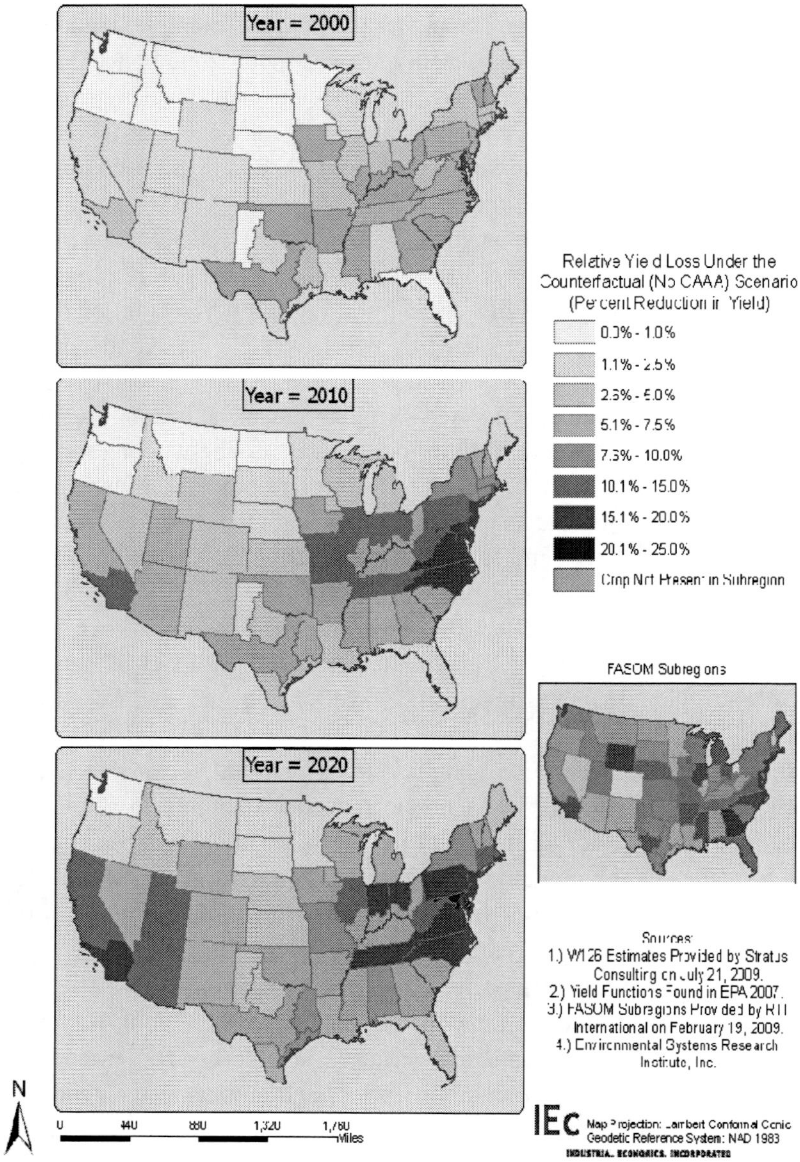

Figure 6-4. Relative Annual Yield Losses in Potatoes Under the Counterfactual (No CAAA) Scenario by Fasom Subregion and Year Based on Subregionalspecific Ozone Concentrations and Growing Periods.

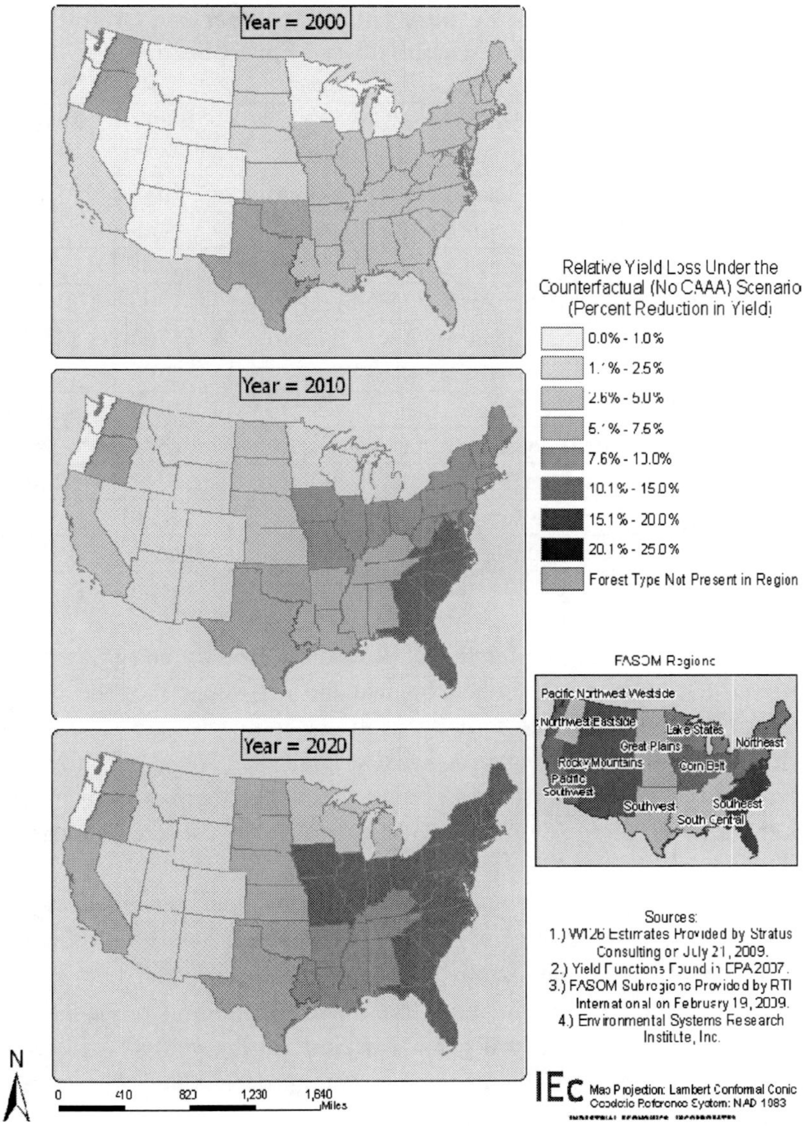

FIGURE 6-5. Relative Annual Yield Losses in Softwood Forest Types Under the Counterfactual (No Caaa) Scenario by Fasom Region and Year Based on Regional-Specific Ozone Concentrations and Growing Periods.

The economic welfare results of the FASOM modeling are presented in Table 6-6. FASOM generates total welfare estimates for the agricultural and forest sectors for each of our scenarios, for each target year, reflecting the sum of total consumer and producer surplus derived from agriculture and forest production. In general, higher ozone concentrations in the *without-CAAA* scenario lead to reduced agricultural and forest productivity, raising prices for these products, which in turn increases producer surplus but reduces consumer surplus by a larger amount. As a result, FASOM estimates the net welfare benefits of the CAAA to be approximately $1 billion in 2000, $5.5 billion in 2010, and $10.7 billion in 2020, increasing over time as the differences in ozone concentrations grows.[86]

Table 6-6. Summary of Fasom Results: Total Consumer and Producer Surplus Values for the Agricultural and Forest Sectors.

VARIABLE	MODEL RUN	2000	2010	2020
Annual Welfare, US Forest Sector	With Clean Air Act ($ billion)	$637	$877	$1426
	Without Clean Air Act ($ billion)	$636	$875	$1426
	Damage Estimate ($ billion)	$1.5	$1.7	$0
	Percent change	0.24%	0.20%	0%
Annual Welfare, US Agriculture Sector	With Clean Air Act ($ billion)	$1706	$1831	$1916
	Without Clean Air Act ($ billion)	$1706	$1828	$1905
	Damage Estimate ($ billion)	−$0.5	$3.8	$10.6
	Percent change	-0.03%	0.21%	0.55%
Annual Welfare, Forest and Agriculture Sector Combined	With Clean Air Act ($ billion)	$2343	$2708	$3341
	Without Clean Air Act ($ billion)	$2242	$2703	$3331
	Damage Estimate ($ billion)	$1.0	$5.5	$10.7
	Percent change	0.05%	0.20%	0.32%

Notes: 1. Results are expressed in year 2006 dollars.

In general, FASOM forecasts a relative shift towards forestry and away from agriculture under the *without-CAAA* scenario, indicating that the net impacts of the ozone effects on forests and agriculture would make forestry relatively more profitable than in the baseline compared with agriculture, resulting in a shift in land use. The model forecasts a sizable increase in cropland in the *without-CAAA* scenario, however there is an even greater decline in pasture as the returns to crop production rise relative to livestock production with higher crop prices.

As noted above, the model suggests that the damages attributed to higher ozone concentrations indicate that producers gain in many cases, while consumers are always substantially worse off with the ozone impacts reducing productivity. The reason that producers often are better off is that most forest and agricultural products have relatively inelastic demands, which means that a general decline in productivity will tend to increase prices by more than the reduction in quantity, increasing revenue and often profits as well. In general, FASOM attributes large price increases in response to the reductions in productivity for these inelastic products, and production declines in the *without-CAAA* scenario for most agricultural commodities, with larger declines in general for those products experiencing larger ozone impacts, and also sizable reductions in exports.

FASOM also is capable of modeling land-use changes in response to the higher ozone concentrations in the *without-CAAA* scenario. The model indicates changes in major land use categories at the national level over time under the ozone impacts scenario, which is leading to a net increase in forest of about 6.1 million acres by the 2020 model period and an increase in cropland of 7.6 million acres by 2020 in response to the productivity declines. At the same time, the model indicates that cropland pasture (high-quality land that is suitable for cropland but is being used as pasture) and pasture (lower-quality land that is not suitable for growing crops without improvement) decline by a total of 12.7 million acres and Conservation Reserve Program (CRP) land decreases by about 1 million acres. The crop experiencing the largest reduction in acreage is soybeans, while there is an increase in wheat acreage and a number of smaller shifts between alternative crops.

Visibility

Air pollution impairs visibility in both residential and recreational settings, and an individual's willingness to pay (WTP) to avoid reductions in visibility differs in these two settings. Benefits of residential visibility relate to the impact of visibility changes on an individual's daily life (e.g., at home, at work, and while engaged in routine recreational activities). Benefits of recreational visibility relate to the impact of visibility changes manifested at parks and wilderness areas that are expected to be experienced by recreational visitors. For the purposes of this analysis, recreational visibility improvements are defined as those that occur specifically in federal Class I areas, and residential visibility improvements are those that occur within the boundaries of Census Metropolitan Statistical Areas (MSAs).

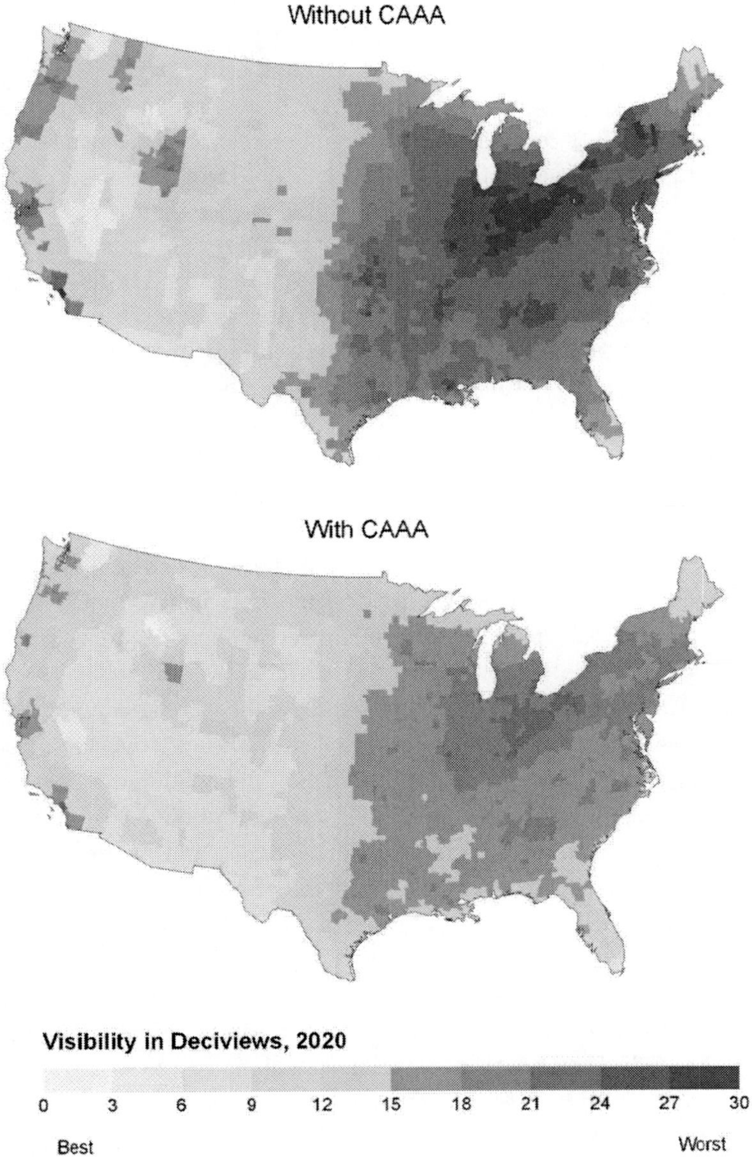

Figure 6-6. Estimated Change in Visibility for with-CAAA Scenario, 1990 to 2020.

We calculate household WTP for improvements in both residential and recreational visibility. We base our calculations on simulations of future visibility conditions at the 36-km grid-cell level, as estimated by EPA's Community Multiscale Air Quality (CMAQ) model. The relationship between a household's WTP and changes in visibility is derived from a number of contingent valuation (CV) studies published in the peer- reviewed economics literature. The approach we apply to estimate the benefit of improvements in recreational visibility is consistent with methods EPA has used in analyses conducted since EPA's First Prospective analysis was completed. In particular, this chapter relies heavily on research completed for the PM NAAQS RIA (U.S. EPA, 2006) for the recreational visibility analysis. Our estimate of the benefit of residential visibility is consistent with methods applied in past analyses as well, but in previous reviews the Council had expressed concerns about residential visibility estimates based on WTP estimates from the McClelland et al. (1991) study. As a result, our estimates in this chapter rely on a new "benefits transfer" estimate of WTP derived from other published sources of residential visibility WTP.

According to the CMAQ simulations, the CAAA has had and will continue to have a substantial effect on visibility in both residential and recreational settings. The visibility data used in this analysis is annual mean visibility data, by county, measured in deciviews.[87] Figure 6-6 depicts the change in visibility (measured in deciviews) over the 30-year time frame, from 1990 to 2020, along the *with-CAAA* scenario. This map shows that, overall, changes in visibility due to the CAAA are greater in the eastern U.S. than the western U.S. Additionally, the largest changes in visibility occur in the Midwestern states. The county level data presented here are the basis for the residential visibility improvements we present below.

Only one existing study provides defensible monetary estimates of the value of recreational visibility (Chestnut and Rowe, 1990b; 1990c). Although the Chestnut and Rowe study is unpublished, it was originally developed as part of the National Acid Precipitation Assessment Program (NAPAP) and, therefore, has been subject to peer- review as part of that program. The Chestnut and Rowe study measures the demand for visibility in Class I areas managed by the National Park Service (NPS) in three broad regions of the country: California, the Southwest, and the Southeast. Respondents in five states were asked about their WTP to protect national parks or NPS-managed wilderness areas within a particular region. The survey used photographs reflecting different visibility levels in the specified recreational areas. The visibility levels in these photographs were later converted to deciviews for the current analysis. The three regions assessed in the study cover 86 of the 156 Class I areas in the United States. Given that national parks and wilderness areas exhibit unique characteristics, it is not clear whether the WTP estimate obtained from the Chestnut and Rowe study can be transferred to other national parks and wilderness areas, without introducing additional uncertainty. As a result, for the primary estimate, we value only those recreational benefits in the areas that were directly analyzed in the original Chestnut and Rowe study.

In the First Prospective analysis, we omitted the results of the benefits estimate for residential visibility from the primary benefits estimate due to technical concerns about the methodology of the study upon which our original calculations were based (McClelland et al., 199 1).[88] There exists a wide range of published, peer-reviewed literature, however, that supports a non-zero value for residential visibility. As a result, we have revised our methodology for valuing residential visibility, and now include these benefits in our overall primary visibility benefits estimate.

For valuing residential visibility improvements, we rely upon a benefits transfer approach that draws upon information from the published Brookshire (1979), Loehman (1984) and Tolley (1986) studies. Each of the studies used provides estimates of household WTP to improve visibility conditions from a status quo visual range to an improved visual range. While uncertainty exists regarding the precision of these older, stated-preference residential valuation studies, we believe their results support the argument that individuals have a non-zero value for residential visibility improvements. The implied annual per-household WTP estimates from these study, for a hypothetical 10-percent improvement, ranges from $14 to $145, with a mean of $69 and median of $53. It is not surprising that such a range of values exists, as the areas of the country covered feature different landscapes and vistas, populations and prevailing visibility conditions.

Fortunately, the three recommended studies provide primary visibility values for a variety of cities throughout the United States: Atlanta, Boston, Chicago, Denver, Los Angeles, Mobile, San Francisco, and Washington D.C. We assign each of the 359 MSAs in the contiguous U.S. a value based on geographic proximity to one of the eight study cities, with two exceptions: 1) We apply the Loehman et al. (1984) value only to the six San Francisco Bay area MSAs, because the study is unique among the three in the manner in which visibility changes were described to respondents (i.e., a distribution of days versus average conditions), and 2) Values associated with Denver are not assigned on the basis of proximity but are instead assigned only to MSAs which meet an elevation range threshold of 1500 meters within the MSA, because one would expect that residents of Denver, with a dramatic view of the Rocky Mountains that is rarely obstructed by trees, would have a greater interest in protecting visibility than a city without a dramatic skyline or nearby mountains.[89]

The primary estimate of benefits of recreational and residential visibility improvements is provided in Table 6-7. The primary estimate for recreational visibility only includes benefits in the original study regions (i.e., California, the Southwest, and the Southeast). The primary estimate for residential visibility includes benefits in all MSAs. In general, benefits to visibility increase over time as visibility improves due to the CAAA. Benefits to residential visibility are approximately three times as large as benefits to recreational visibility.

Table 6-7. Primary Estimate of Benefits to Visibility (Billion 2006$).

	2000 BENEFITS	2010 BENEFITS	2020 BENEFITS
Recreational Benefits	$3.3	$8.6	$19
Residential Benefits	$11	$25	$48
Total Benefits	$14	$34	$67

Residential visibility benefits are driven by population and visibility improvements. Overall, benefits are greater in the East. This is due in part to greater population levels as well as greater visibility improvements. Benefits are also very high in California due to the state's large population and visibility improvements, especially in and around Los Angeles and San Francisco. Residential visibility is also dependent upon the WTP value applied. Much of the West uses the WTP value for Denver, which is highest WTP value being widely applied. Yet, the West still has lower overall benefits to residential visibility.[90] This impact shows that the effect of population and visibility improvement dominates the effect of the WTP value applied.

Materials Damage

Since the mid-19th century air pollution has been suspected of accelerating the degradation of natural and man-made materials that are exposed to the outdoor environment. Concern over the effect of pollutants on materials has mainly been directed towards the economic consequences of damage to materials used in construction, but aesthetic damage to historic buildings and monuments is also a concern. Wet and dry acidic deposition, alone or combined with other air pollutants, contribute to the increased rate of materials damage. Acidic deposition has been shown to have an effect on materials including zinc/galvanized steel and other metal, carbonate stone (as monuments and building facings), and surface coatings (paints) (NAPAP, 1991).

Table 6-8. Materials Damage Effects.

POLLUTANT	QUANTIFIED EFFECTS—DAMAGE TO:	UNQUANTIFIED EFFECTS[a]—DAMAGE TO:
Sulfur oxides	Infrastructural materials – galvanized and painted carbon steel Commercial buildings – carbonate stone, metal, and painted wood surfaces Residential buildings – carbonate stone, metal, and painted wood surfaces	Monuments – carbonate stone and metal Structural aesthetics Automotive finishes – painted metal
Hydrogen ion and nitrogen oxides		Infrastructural materials – galvanized and painted carbon steel Zinc-based metal products, such as galvanized steel Commercial and residential buildings – carbonate stone, metal, and wood surfaces Monuments – carbonate stone and metal Structural aesthetics Automotive finishes – painted metal
Carbon dioxide		Zinc-based metal products, such as galvanized steel
Formaldehyde		Zinc-based metal products, such as galvanized steel
Particulate matter		Household cleanliness (i.e., household soiling)
Ozone		Rubber products (e.g., tires)

a The categorization of unquantified effects is not exhaustive.

Metal structures are usually coated by alkaline corrosion product layers and thus are subject to increased corrosion by acidic deposition. In addition, research has demonstrated that iron, copper, and aluminum based products are subject to increased corrosion due to pollution, in particular SO_2 (NAPAP, 1991), that acidic deposition accelerates the rate of erosion of carbonate stone (marble and limestone), and that acidic deposition has numerous negative effects on painted wood and, in general, increases the weathering rate. This analysis focuses on quantifying the impact of sulfur dioxide deposition on exterior building and infrastructural materials including carbonate stone, galvanized steel, carbon steel, and painted wood, as outlined Table 6-8 above.

This analysis applies the Air Pollution Emissions Experiments and Policy (APEEP) analysis model, described in Muller and Mendelsohn (2007, 2009), to link SO_2 emissions to ambient SO_2 levels. Using emission inputs, the air quality model in APEEP forecasts seasonal and annual average county concentrations for SO_2, amongst other pollutants.[91] As reported in Muller and Mendelsohn (2007) and detailed in the supporting online material for that publication, APEEP's air quality modeling has been statistically tested and calibrated against the predictions generated by the Community Multi-scale Air Quality Model (CMAQ), using 1996 emissions data and a CMAQ run for 1996 conditions. Muller and Mendelsohn (2007) also report comparisons of APEEP's results with available monitor data for this period. The results for the SO_2 air quality component used in these materials damage calculations appear to suggest good agreement for APEEP for concentrations near the mean, but APEEP appears to overpredict SO_2 concentrations for high-end concentrations. Overall, however, it is important to note that APEEP is designed to be a fast-running alternative to CMAQ for use in an integrated assessment model – the air quality component of APEEP is a statistical representation of relations that are accomplished in a far more sophisticated manner in CMAQ.

The remaining general steps in the process of estimating materials damage effects are as follows:

- **Develop a national inventory of sensitive materials**. A key piece of information needed to apply the appropriate materials damage concentration-response functions is the existing materials inventories. This analysis estimates the inventory of four exterior building and infrastructural materials in each county in the lower 48 states, including carbonate stone, galvanized steel, carbon steel, and painted wood surfaces.
- **Derive concentration-response functions that relate material mass loss to ambient SO_2.** Dose-response functions for man-made materials damages are obtained from two sources; the NAPAP studies (Atteraas, Haagenrud, 1982; Haynie, 1986) and from the International Cooperative Programme on Effects on Materials (ICP, 1998).
- **Estimate the value of lost materials.** Materials damage is valued as the cost of future materials maintenance activities. The accelerated rate of materials decay due to pollution exposure increases the frequency of regularly scheduled future maintenance activities. The change in the present value of the maintenance schedules extending into the future constitutes the monetary impact of an emission change on materials damage.

Table 6-9. Benefits of Reduced Materials Damage due to CAAA Programs.

EPA REGION	VALUATION (THOUSAND 2006$)		
	2000	2010	2020
1: CT, ME, MA, NH, RI, VT	$720	$2,100	$2,100
2: NY, NY	$9,000	$10,000	$12,000
3: DE, DC, MD, PA, VA, WV	$9,400	$19,000	$23,000
4: AL, FL, GA, KY, MS, NC, SC, TN	$8,400	$16,000	$21,000
5: IL, IN, MI, MN, OH, WI	$26,000	$38,000	$38,000
6: AR, LA, NM, OK, TX	$2,200	$4,000	$7,300
7: IA, KS, MO, NE	$2,000	$1,600	$1,600
8: CO, MT, ND, SD, UT, WY	$400	$570	$730
9: AZ, CA, NV	-$100	$490	$640
10: ID, OR, WA	$340	$510	$560
Total	**$58,000**	**$93,000**	**$110,000**

Notes: Results are rounded to two significant figures. Totals may not sum due to rounding.

Table 6-9 summarizes the benefits of reduced materials damage attributed to CAAA programs in 2000, 2010, and 2020. Benefits are given by EPA region. Although the total benefits are relatively small compared to other categories of effect, the benefits of CAAA programs to materials damage increase over time as we would expect. The spatial distribution of the benefits is primarily owing to the distribution of the materials inventory and SO_2 exposure. The effect of SO_2 exposure is a more important driver of results than the inventory. For example, the benefits in Region 5 are approximately twice as large as those in any other EPA region. This is due to the significant decrease in SO_2 exposure associated with the CAAA in this region.

Adirondack Case Study Results

The Project Team was encouraged to consider case study analysis of a set of ecological effects for which national analyses might not be feasible, owing to lack of available data or methods. EPA chose to conduct a case study in the Adirondack region of New York State, focusing on two ecological service flows that provide benefits in terms of both ecosystem health and economic terms: (1) acidification of surface waters and (2) reduced yields of commercial timber. The Adirondack region of New York may exhibit the most severe ecological impacts from acidic deposition of any region in North America – acid deposition is the main cause of both of the effects we studied.[92] Adirondack Park is a State Park comprising 5,821,183 acres of State and privately owned land in upstate New York and is nearly a 100 by 100 mile box of land, intersecting fourteen counties. The Park was created in 1892 through an amendment to the State constitution, with the purpose of forest and natural resource conservation. Federal programs addressing air pollution have been particularly beneficial to the region as, due to its location downwind of the highly industrialized Ohio River Valley, most of the acid deposition in the region originates from out of state. In addition to its status as a region of particular sensitivity to lake acidification and with some existing research on the effects of air pollutants on forest growth, the Adirondack Region was selected as a setting for this case study due to the existence of a regional economic random utility model describing recreational fishing behavior.

Lake Acidification in the Adirondacks

Surface waters, such as lakes and streams, may be the most susceptible systems to acidic deposition as they collect acidic precipitation not only from direct deposition on their surfaces but also in the form of runoff from their entire watershed. Acid accumulates in surface waters via three main pathways:

- precipitation, or wet deposition, in which pollutants are dissolved in rain or snow;
- dry deposition, or direct deposition of gases and particles on surfaces; and
- cloud-water deposition, involving material dissolved in cloud droplets and deposited on vegetation.[93]

As acids accumulate, ecosystems gradually lose the ability to buffer them, resulting in changes to ecosystem structure and function. Acidification of the surface water affects the trophic structure of water contributing to declines in the abundance of zooplankton, macroinvertebrates, and fish.[94]

The ecological service flow affected by lake acidification that is most amenable to economic analysis is recreational fishing. Extensive research exists focused on both the effects of lake acidification on fisheries and on individuals' willingness to pay to avoid reductions in the quality or quantity of recreational fishing opportunities. This analysis employs the following general steps to quantify the benefits of reduced lake acidification on recreational fishing in the Adirondacks. A conceptual model depicting the analytic steps in terms of inputs, outputs, and ecological and economic models is provided in Figure 6-8.

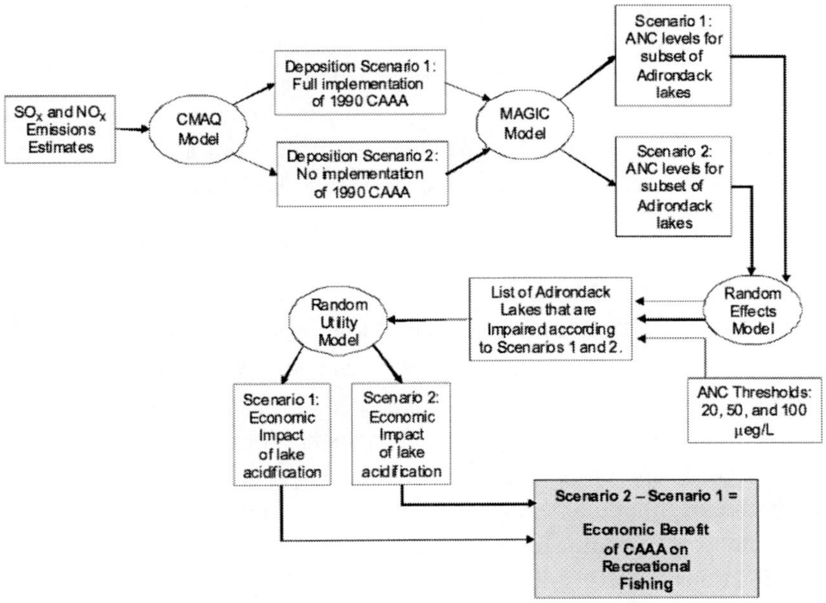

Figure 6-8. Conceptual Approach to Estimating the Economic Benefits of Reduced Acidification on Adirondack Lakes.

- **Forecast lake acidification levels consistent with the *with-CAAA* and *without-CAAA* scenarios.** EPA generated estimates of acidic deposition at a 36-kilometer grid cell level across the Adirondack region using the CMAQ model. We then implemented an ecological model, the Model of Acidification of Groundwater in Catchments (MAGIC), to simulate the transport of the acidic deposition through the hydrological and terrestrial ecosystems and forecast acidification levels in a subset of Adirondack lakes.
- **Extrapolate results of the ecological model within the Adirondacks region.** We developed a random effects model to explain the relationship between acidification of lakes and their specific site characteristics.
- **Apply ANC thresholds to classify lakes as either "fishable" or "impaired".** Fishable lakes are those for which water quality is not deteriorated to an extent which limits recreational fishing. Impaired lakes' water quality is deteriorated so as to reduce fish populations and preclude recreational fishing. Lakes are defined as either fishable or impaired based on identified ANC thresholds. As uncertainty exists regarding the ANC threshold at which effects are experienced, this analysis considers three separate thresholds below which lakes are considered impaired.
- **Apply an economic random utility model (RUM) to quantify economic benefits of the CAAA in terms of recreational fishing in the Adirondack region.** We employ a RUM that was developed to account for fishing site choices made by recreational fishers based on attributes of sites specifically in the Adirondack region. The difference in economic welfare values between the value of fishable (i.e., not impaired) lakes in the *with-CAAA* scenario and the *without-CAAA* scenario represents the benefits to recreational fishing in the Adirondack region associated with the CAAA.

Table 6-10 summarizes the results of this analysis. Present value cumulative benefits are provided for 2000, 2010, and 2020, assuming a five percent discount rate. Single year undiscounted benefits are also given for each year. Undiscounted single year benefits increase over time but the benefits do not follow any particular trend across alternative threshold assumptions. It should be noted that benefits in each year and under each threshold assumption reflect a different subset of lakes. Therefore, benefits are not expected to follow any particular trend across years or threshold assumptions.

Commercial Timber in the Adirondacks

Reductions in NO_x and SO_x emissions due to the implementation of the CAAA are also believed to reduce forest soil acidity. Reductions in soil acidity have been shown by scientists to increase tree growth and improve overall forest health. Such changes in forest growth and health would have a positive effect on the timber industry within Adirondack Park, potentially increasing the frequency and/or the volume of timber harvests.

Quantifying the magnitude of these benefits requires a function to translate varying levels of soil acidity into corresponding tree growth productivity. Unfortunately, species-specific dose-response functions relating soil acidity levels with changes in tree growth in Adirondack Park are not available. Our analysis instead characterized the existing timber industry in Adirondack Park in terms of the types of tree species present, wood products harvested, extent of timber harvest activities, and the overall value of timber harvests within the Park. We then estimated changes in percent base saturation (a measure of soil acidity) due to the implementation of the CAAA across the Park from 1990 to 2050, focusing on soil acidity differences in areas subject to commercial timber activity. Specifically, changes in percent base saturation levels in timber harvest areas were mapped in relation to potential changes in the growth and health of tree species present in these areas and the likely effects of altered tree growth and health on timber harvest rates and volumes. In addition, we provide some perspective on the potential order of magnitude of benefits of the CAAA on the timber industry in the Adirondacks, summarizing existing, relevant research.

We used estimates of soil percent base saturation levels for 1990, 2000, 2010, 2020, and 2050 with and without the CAAA to characterize the effect on Adirondack forests.[95] Percent base saturation is the proportion of cation exchange sites (exchange sites are areas on soil particles where ions may be adsorbed) occupied by basic cations (Ca^{2+}, Mg^{2+}, K^{+}, and Na^{+}). These basic cations buffer the soil by inhibiting the adsorption of H^{+} ions. Thus, percent base saturation is a measure of the soil's buffering capacity. High percent base saturation levels indicate large buffering capacity and low soil acidity levels, while low percent base saturation levels indicate the converse. Percent base saturation point estimates were generated using the same Model of Acidification of Groundwater in Catchments (MAGIC) as used in the lake acidification analysis described above.

Table 6-10. Summary of Annual and Cumulative Estimated Benefits to Recreational Fishing in the Adirondack Region (Million 2006$).

YEAR	ANC THRESHOLD ASSUMPTION FOR DEFINING "FISHABLE" LAKES	ADIRONDACK REGION	
		SINGLE YEAR UNDISCOUNTED	CUMULATIVE FIVE PERCENT DISCOUNT RATE
2000	20	$7	$62
	50	$7	$57
	100	$5	$44
2010	20	$8	$143
	50	$8	$132
	100	$6	$101
2020	20	$9	$197
	50	$8	$182
	100	$6	$136

Note: 1) Cumulative benefits in year 2000 are the cumulative benefits to recreational fishing of implementing the CAAA from 1990 to 2000. Similarly, cumulative benefits in 2010 are cumulative from 1990 to 2010 and cumulative benefits in 2020 are cumulative from 1990 to 2020. The single year undiscounted benefits are the benefits to recreation fishing of implementing CAAA in that year (2000, 2010, or 2020).

2) Benefits in this case study are evaluated from 1990 (the year of the passage of the CAAA) to 2050 (the forecast horizon for the lake ANC levels with and without the CAAA). The benefits in this table are presented for years 2000, 2010, and 2020, however, to be consistent with the benefits as calculated in the broader cost-benefit analysis of the CAAA.

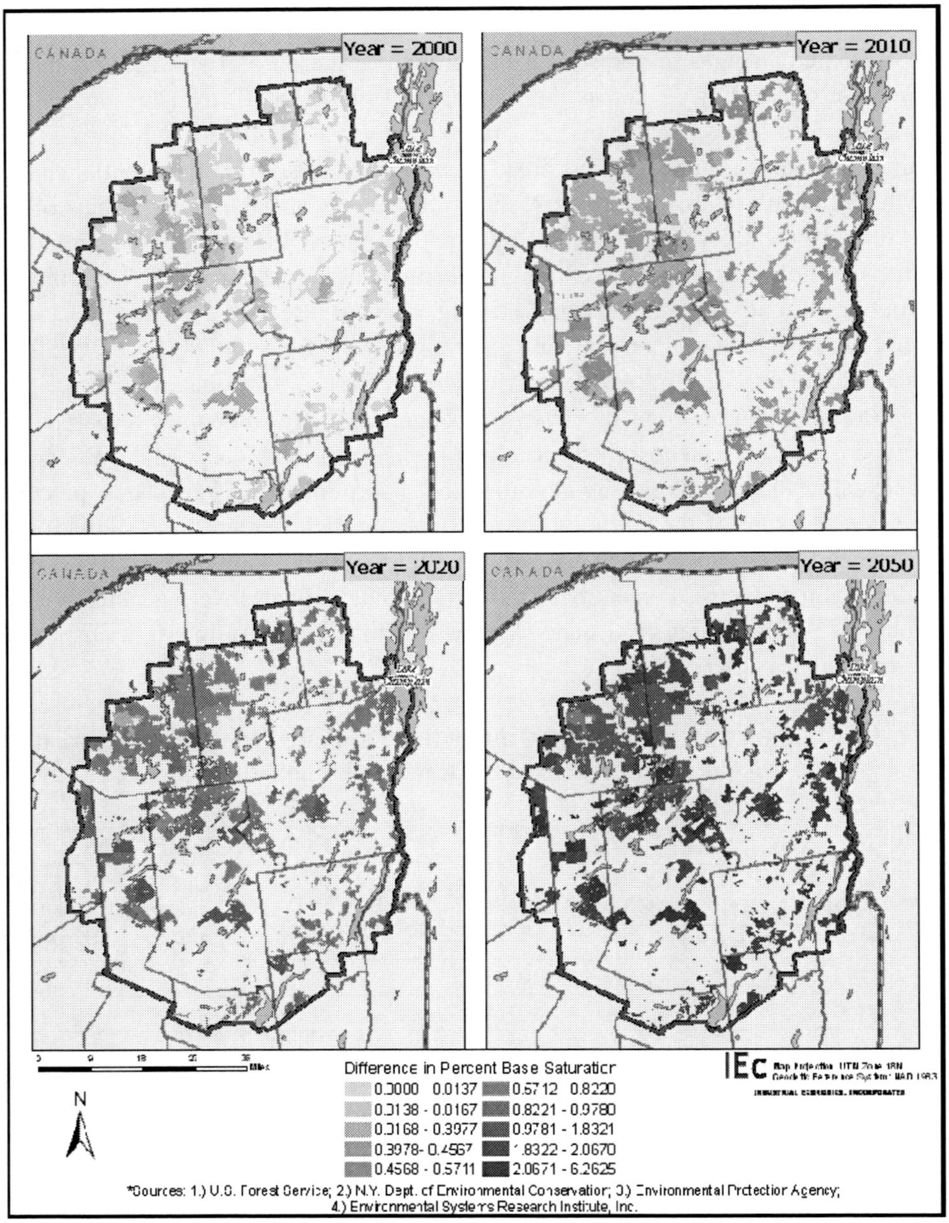

Figure 6-9. Differences in Percent Base Saturation Values with and Without the CAAA in Forested Resource Management Areas in Adirondack Park. [96, 97]

Figure 6-9 presents differences in percent base saturation levels with and without the CAAA specifically within the timber harvest areas of the Park by year. There is a clear temporal trend in the difference in percent base saturation levels with and without the CAAA. Specifically, differences between percent base saturation levels with the CAAA as compared to without the CAAA increase in each year in the analysis. However, there is little spatial variability in percent base saturation differences within individual years. The lack of spatial variability becomes more pronounced as time goes on, so that by 2050 the difference in percent base saturation is between 2.07 and 6.26 percent in almost all forested resource management areas in the Park. The lack of spatial variability makes sense given the relatively small geographic scope considered in this analysis. The minor spatial variation in percent base saturation differences exhibited in 2000 and 2010 is most likely related to microhabitat factors (i.e., different soil types and differing precipitation levels).

Also of importance to this analysis is the magnitude of the increase in percent base saturation levels in relation to specific forest types within resource management areas. We focused on six forest types (sugar maple/beech/yellow birch, red maple/upland, spruce/fir, eastern hemlock, eastern white pine, and paper birch) that are prevalent in the Park relative to other forest types and contain tree species of commercial value. Table 6-11 presents the area-weighted mean increase in percent base saturation levels in these forest types per year. Of the forest types of interest, the paper birch forest type experiences the greatest increase in percent base saturation due to the CAAA, followed by the eastern hemlock and the sugar maple/beech/yellow birch forest types.

The area-weighted increase in percent base saturation levels in sugar maple/beech/yellow birch forests is in line with increases in percent base saturation levels in other forest types in Adirondack Park. This is an important point given the prevalence of sugar maple in this forest type, and the fact that sugar maple is an economically important tree species in the Park. Although dose-response functions, which would allow for estimates of growth increases in sugar maples due to increased base saturation levels, do not exist, several studies have estimated changes in sugar maple growth due to increases in soil acidity stemming from elevated nitrogen and/or sulfur deposition.[98] Changes in harvest volumes comparable to those seen in those existing studies might lead to annual wood harvest benefits of roughly $1 million to $1.5 million annually, based on the total stumpage values for sugar maple pulpwood/chip wood we estimate for the region.[99] Whether sugar maple growth rate changes would mirror those reported in either of these studies, however, is uncertain due to the lack of an established functional relationship.

Table 6-11. Area-Weighted Mean Differences in Percent Base Saturation Values with and without the CAAA in Forest Types of Interest.

FOREST TYPE	AREA-WEIGHTED DIFFERENCE IN PERCENT BASE SATURATION			
	2000	2010	2020	2050
Sugar Maple/Beech/Yellow Birch	0.023	0.414	0.820	1.899
Red Maple/Upland	0.025	0.377	0.758	1.755
Spruce/Fir	0.028	0.361	0.736	1.702
Eastern Hemlock	0.028	0.413	0.827	1.908
Eastern White Pine	0.018	0.419	0.814	1.882
Paper Birch	0.018	0.457	0.891	2.069
Other Forest Types	0.015	0.429	0.829	1.918

Nonetheless, we expect that all tree species in the Park would benefit, in terms of increased stand growth and vigor, from increased percent base saturation levels. In some cases, increases in growth may allow for both more frequent and larger timber harvests (i.e., more frequent timber harvests removing larger volumes of wood). Improved forest health may also provide the added benefit of increasing the resiliency of forest stands and limiting damage caused by disturbance events.

Uncertainty in Ecological and Other Welfare Benefits

As noted above, limitations in the available methods and data mean that the benefits assessment in this report does not represent a comprehensive estimate of the economic benefits of the CAAA. Moreover, the potential magnitude of long-term economic impacts of ecological damages mitigated by the CAAA suggests that great care must be taken to consider those ecosystem impacts that are not quantified here. Significant future analytical work and basic ecological and economic research is needed to build a sufficient base of knowledge and data to support an adequate assessment of ecological benefits. For the current analysis, this incomplete coverage of effects represents the greatest source of uncertainty in the ecological assessment. This and other key uncertainties are summarized in Table 6-12 below.

Table 6-12. Key Uncertainties Associated with Ecological Effects Estimation.

POTENTIAL SOURCE OF ERROR	DIRECTION OF POTENTIAL BIAS FOR NET BENEFITS	LIKELY SIGNIFICANCE RELATIVE TO KEY UNCERTAINTIES ON NET BENEFITS ESTIMATE*
Incomplete coverage of ecological effects identified in existing literature, including the inability to adequately discern the role of air pollution in multiple stressor effects on ecosystems. Examples of categories of potential ecological effects for which benefits are not quantified include: reduced eutrophication of estuaries, reduced acidification of soils, reduced bioaccumulation of mercury and dioxins in the food chain.	Underestimate	Potentially major. The extent of unquantified and unmonetized benefits is largely unknown, but the available evidence suggests the impact of air pollutants on ecological systems may be widespread and significant.
Incomplete geographic scope of recreational fishing benefits associated with reduced lake acidification analysis due to case study approach.	Underestimate	Probably minor. As a case study limited to the Adirondack region of New York State, the estimated benefits to recreational fishing reflect only a portion of the overall benefits of reduced acidification on this service flow, but based on the magnitude of effects in the Adirondacks the national estimate is nonetheless likely be less than five percent of total benefits.
Incomplete assessment of long-term bioaccumulative	Underestimate	Potentially major. Little is currently known about the

Table 6-12. (Continued)

POTENTIAL SOURCE OF ERROR	DIRECTION OF POTENTIAL BIAS FOR NET BENEFITS	LIKELY SIGNIFICANCE RELATIVE TO KEY UNCERTAINTIES ON NET BENEFITS ESTIMATE*
and persistent effects of air pollutants.		longerterm effects associated with the accumulation of toxins in ecosystems. What is known suggests the potential for major impacts. Future research into the potential for threshold effects is necessary to establish the ultimate significance of this factor.
Omission of the effects of nitrogen deposition as a nutrient with beneficial effects.	Overestimate	Probably minor. Although nitrogen does have beneficial effects as a nutrient in a wide range of ecological systems, nitrogen in excess also has significant and in some cases persistent detrimental effects that are also not adeq-uately reflected in the analysis.
Use of CMAQ model to estimate air pollutant deposition levels.	Unable to determine. As part of a performance evaluation of CMAQ, EPA compared model predictions for some forms of deposition relevant to this analysis (wet SO_2, NO_x and ammonium) to observed deposition data).** The evaluation indicated that CMAQ overpredicted some forms of deposition and underpredicted others. The relative accuracy of the model's predictions varied seasonally and geographically.	Probably minor. The Adirondack lake acidification analysis uses deposition estimates as inputs, but they are calibrated to lake-level monitoring data, and the monetized benefits estimates for that component are a small part of the overall net benefits. We also use the CMAQ deposition estimates to generate maps that highlight the relative distribution of deposition for various air pollutants across the U.S. With respect to net impacts, the extent to which the forms of deposition and geographic areas that are overpredicted in the model are offset by those that are underpredicted is unknown.

* The classification of each potential source of error reflects the best judgment of the section 812 Project Team. The Project Team assigns a classification of "potentially major" if a plausible alternative assumption or approach could influence the overall monetary benefit estimate by approximately five percent or more; if an alternative assumption or approach is likely to change the total benefit estimate by less than five percent, the Project Team assigns a classification of "probably minor."

** See U.S. EPA, Office of Air Quality Planning and Standards, Emissions Analysis and Monitoring Division, Air Quality Modeling Group. CMAQ Model Performance Evaluation Report for 2001: Updated March 2005. CAIR Docket OAR-2005-0053-2149.

In general, our analysis focuses on more acute and readily observable effects. Chronic ecological effects of air pollutants, on the other hand, may be poorly understood, difficult to observe, or difficult to discern from other influences on dynamic ecosystems. Disruptions that may seem inconsequential in the short-term, however, can have hidden, long-term effects through a series of interrelationships that can be difficult or impossible to observe, quantify, and model. This factor suggests that many of our qualitative and quantitative results may underestimate the overall, long-term effects of pollutants on ecological systems and resources.

7. Comparison of Benefits and Costs

In this chapter we present our summary of the primary estimates of monetized benefits of the CAAA from 1990 to 2020, compare the benefits estimates with the corresponding costs, and explore some of the major sources of uncertainty in the benefits estimates, including a summary of outcomes using alternative assumptions from those employed in the primary analysis.

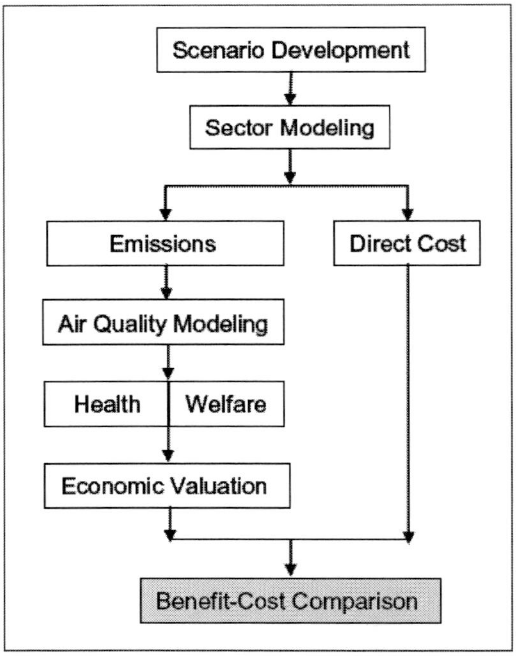

The overall conclusion of our analysis is that the benefits of the CAAA *substantially* exceed its costs. Furthermore, the results of the uncertainty analysis imply that it is extremely unlikely that the monetized benefits of the CAAA over the 1990 to 2020 period could be less than its costs. The central benefits estimate exceeds costs by a factor of more than 30 to one, whether we are looking at annual or present value measures. By our measures, the programs associated with the 1990 Clean Air Act Amendments have been, and will likely continue to be, a very good investment.

Aggregating Benefit Estimates

Our primary estimates of the monetized economic benefits for the 1990 to 2020 period derive from two types of analyses: (1) the analysis of changes in human health effects associated with reduced exposures to criteria pollutants and the valuation of these changes, summarized and described in Chapter 5; and (2) the analysis of monetized ecological and other welfare benefits (e.g., visibility), described in Chapter 6.[100] We measure the benefits and present the results from these analyses in slightly different ways, in part because they derive from different tools. The main differences have to do with ti manner in which we conduct uncertainty analyses, as outlined below.

Although there are some differences in these two types of benefits analysis, in both cases we generate annual estimates of benefits that result from a single set of emissions and air quality modeling scenarios for the three target years of the study: 2000, 2010, and 2020. The consistent use of scenarios across all the benefit and cost analyses allows us to aggregate and directly compare monetized benefits estimates to the estimates of costs incurred in the target years. In some cases, we need to apply a discount rate to compare benefits to costs; for example, we model the effect of particulate matter on premature mortality to occur over a period of twenty years from the time of exposure, even though the costs to achieve that benefit are incurred at the time of the initial exposure change. In this case, we have accounted for the incidence of premature mortality over the assumed lag period, and discounted the valuation of this effect back to the target year. Some ecological effects, such as the effects of acid deposition on Adirondack lakes, also occur with a lag – again, we use a discounting procedure to standardize the benefits results for these estimates.

The annual estimates for the three target years also provide an indication of the trend in benefits we project will accrue over the 30-year study period. To generate a cumulative measure of benefits over the full 30-year period, however, we must make an assumption about the level of benefits that would be realized in the years between the target years. We interpolate these values, assuming a trend in benefits accrual that roughly matches th trend in emission reductions for PM precursors. Basing our estimate of the benefits trajectory on PM precursor reductions acknowledges that the majority of monetized benefits, including health and visibility, are attributable to reductions in ambient particulate matter.

The distribution of estimates we generate for the monetized benefits of human health effects incorporates both the quantified uncertainty associated with each of the health effect estimates and the quantified uncertainty associated with the corresponding economic valuation strategy. Quantitative estimates of uncertainties in earlier steps of th analysis (i.e., emissions and air quality changes) could not be developed adequately and are therefore not applied in the present study. As a result, the range of estimates for monetized benefits presented in this chapter, from the primary low estimate to the primary high estimate, is narrower than would be expected with a complete accounting o the uncertainties in all analytical components.[101]

In the health benefits analyses we estimate, for each endpoint-pollutant combination, distributions of values for both the key parameter of the concentration-response function and the valuation coefficients. We combine these distributions by using a computerized, statistical aggregation technique to estimate the mean of the monetized benefit estimate for each endpoint-pollutant combination and to characterize the uncertainty surrounding each estimate.[102]

The ecological and welfare results are not currently amenable to the same type of uncertainty analysis. The modeling procedures for estimating the effects of sulfur and nitrogen deposition in acidifying lakes, the effects of ozone in reducing timber and agricultural production, and the effects of particulate matter on visibility are all subject to uncertainty, but they require substantial resources simply to develop single point estimates. We describe key uncertainties in these estimation procedures qualitatively in Chapter 6, with some limited sensitivity analyses also presented to characterize the effect of key assumptions. The sources of uncertainty in these estimates, however, cannot as easily be disaggregated among physical effects modeling and valuation components, and they have not been assessed with the BenMAP model used for health benefits uncertainty analysis. As a result, we cannot reliably develop an aggregate estimate of the uncertainty in the sum of health and welfare benefits estimates.

Annual Benefits Estimates

We present the results of our aggregation of primary annual health benefits estimates for the CAAA in Figure 7-1 below. The figure provides a characterization of both the primary central estimate and the range of values generated by the aggregation procedure described above, for each of the three target years of the analysis (2000, 2010, and 2020). The Primary High estimate corresponds to the 95th percentile value from the health benefits aggregation, and the Primary Low estimate corresponds to the 5th percentile value. The total benefits estimates are substantial; for example, the Primary Central estimate in 2020 is $2.0 trillion.

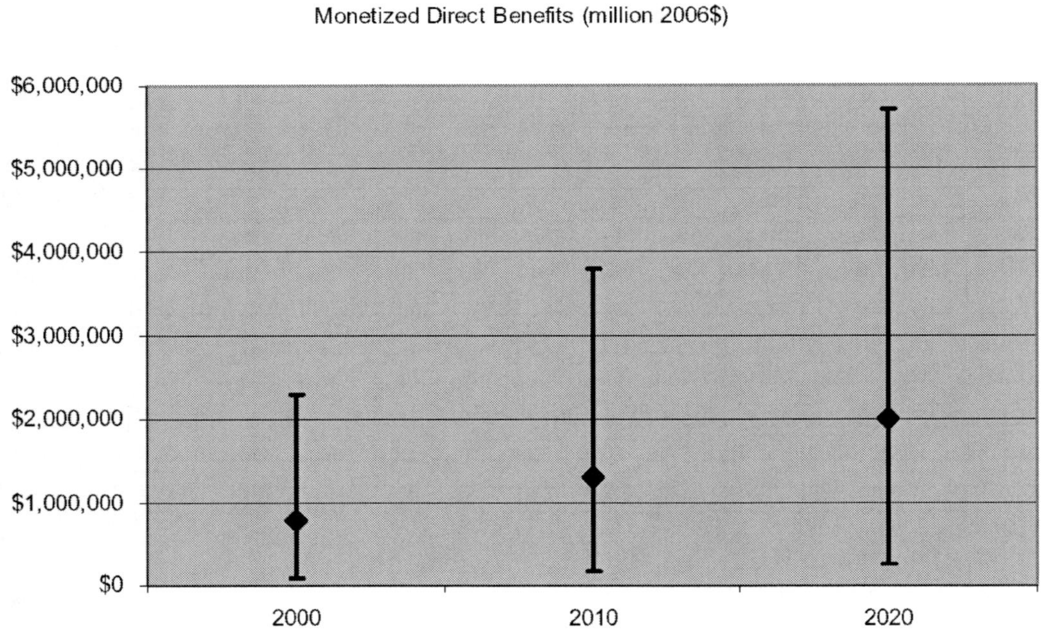

Figure 7-1. Annual Monetized Benefits in 2000, 2010 and 2020.

Table 7-1 shows the detailed breakdown of benefits estimates for 2000, 2010, and 2020. As shown in the table, $1.7 trillion of the $2.0 trillion total benefit estimate in 2020, or 85 percent, is attributable to reductions in premature mortality associated with reductions in ambient particulate matter. The remaining benefits are roughly equally divided among three broad categories of benefits: avoided premature mortality associated with ozone exposure; avoided morbidity, the largest component of which is avoided acute myocardial infarctions and avoided chronic bronchitis; and avoided ecological and other welfare benefits, the largest component of which is improved visibility. Because of the aggregation procedure used, and because we round all intermediate results to two significant digits for presentation purposes, the columns of Table 7-1 may not sum to the total estimate presented in the last row.

Table 7-1. Summary of Mean Primary Annual Benefits Results.

BENEFIT CATEGORY	ANNUAL MONETIZED BENEFITS (MILLION 2006$) BY TARGET YEAR			NOTES
	2000	2010	2020	
Health Effects				
PM Mortality PM Morbidity Ozone Mortality Ozone Morbidity	$710,000 $27,000 $10,000 $420	$1,200,000 $46,000 $33,000 $1,300	$1,700,000 $68,000 $55,000 $2,100	• PM mortality estimates based on Weibull distribution derived from Pope et. Al (2002) and Laden et al., 2006. • Ozone mortality estimates based on pooled function
Subtotal Health Effects	$750,000	$1,300,000	$1,900,000	
Visibility				
Recreational Residential	$3,300 $11,000	$8,600 $25,000	$19,000 $48,000	Recreational visibility only includes benefits in the regions analyzed in Chestnut and Rowe, 1990 (i.e., California, the Southwest, and the Southeast).
Subtotal Visibility	$14,000	$34,000	$67,000	
Agricultural and Forest Productivity	$1,000	$5,500	$11,000	
Materials Damage	$58	$93	$110	
Ecological	$6.9	$7.5	$8.2	Reduced lake acidification benefits to recreational fishing.
Total: all categories	$770,000	$1,300,000	$2,000,000	

Note: See Chapters 5 and 6 of this report for detailed results summaries. Values presented are means from results reported as distributions. Estimates presented with two significant figures.

Examination of the emissions and aggregate exposure estimates suggests that most of these benefits can be attributed to air quality improvements that result from CAAA implementation, relative to conditions as they were in 1990, before the CAAA, rather than from avoiding degradation of air quality that might have occurred without the CAAA. FOR EXAMPLE, WE ESTIMATE THAT EMISSIONS OF NOX, SO_2, AND VOCS, THREE OF THE most important PM and ozone precursors, would have grown just over 20 percent from 1990 to 2020 in the *without-CAAA* scenario, which corresponds to an annual growth rate of about 0.65 percent. We also estimate that $PM_{2.5}$ emissions would have grown somewhat slower, at 0.5 percent annually, without the CAAA. Reductions along the with-CAAA scenario over the same period, however, were more than 60 percent for SO_2 and NO_x, a reduction of roughly 3 percent per year, and were roughly 45 percent for VOCs, a reduction of about 2 percent per year. As a result, about 75 percent of the difference in emissions that we estimate would occur by 2020 between the *with-CAAA* and *without-CAAA* scenario can be attributed to reductions in emissions relative to those in 1990.

$PM_{2.5}$ exposure estimates also support the conclusion that more of the benefit in 2020 can be attributed to air quality improvements from implementing CAAA programs than to preventing degradation in air quality that might have resulted in the *without-CAAA* case. Although we did not estimate 1990 air quality using the CMAQ/MATS system described in Chapter 4, and the $PM_{2.5}$ monitor network was very sparse in 1990, there was an extensive PM_{10} monitor network at that time. Using PM_{10} monitor data and regional $PM_{2.5}/PM_{10}$ ratio estimates from the 1996 Particulate Matter Criteria Document, we estimated population weighted average exposure to $PM_{2.5}$ in 1990 of 19.0 $\mu g/m^3$. In addition, using the CMAQ/MATS system, we estimate population-weighted average exposure to $PM_{2.5}$ along the *without-CAAA* scenario is about 17 $\mu g/m^3$ in 2000, and increases to 17.7 $\mu g/m^3$ and 19.2 $\mu g/m^3$ in 2010 and 2020. Along the *with-CAAA* scenario, population weighted average exposure to $PM_{2.5}$ is 12.2 $\mu g/m^3$ in 2000, and declines to 10.9 $\mu g/m^3$ in 2010, and 10.5 $\mu g/m^3$ in 2020. In the *without-CAAA* scenario some improvements in air quality occurred from 1990 to 2000 as a result of the continuing effect of the pre-1990 Clean Air Act requirements already on the books, but after 2000 the *without-CAAA* scenario shows deterioration of air quality through 2020.

As shown in Table 7-2, there is considerable uncertainty in the estimates of health benefits. As described above, the health benefit uncertainty analysis is based on underlying statistical uncertainties in the concentration-response and valuation coefficients. The low estimates are approximately an order of magnitude less than the central estimate; the high estimate is three times the central estimate. Uncertainty analyses for non-health benefits were not developed, but as they constitute only about five percent of the central estimate, their contribution to the overall uncertainty in benefits estimates is likely to be proportionately small.

Aggregate Monetized Benefits

As discussed earlier in this chapter, we interpolate benefit estimates between target years and then aggregate the resulting annual estimates across the entire 1990 to 2020 period of the study to yield a present discounted value of total aggregate benefits for the period. In this section we present the results of the aggregation.

Table 7-2. Distribution of Primary Annual Benefits Results for 2020.

BENEFIT CATEGORY	PRIMARY ANNUAL MONETIZED BENEFITS FOR 2020 (MILLION 2006$)			NOTES
	LOW	CENTRAL	HIGH	
Health Effects				
PM Mortality PM Morbidity Ozone Mortality Ozone Morbidity	$170,000 $17,000 $3,200 $780	$1,700,000 $68,000 $55,000 $2,100	$5,300,000 $190,000 $170,000 $3,600	Low and high are 5th and 95th percentile estimates from health benefits uncertainty analysis
Subtotal Health Effects	$190,000	$1,900,000	$5,700,000	
Visibility				
Recreational Residential		$19,000 $48,000		Only central estimates were developed
Subtotal Visibility		$67,000		
Agricultural and Forest Productivity		$11,000		
Materials Damage		$110		Only central estimates were developed
Ecological		$8.2		Reduced lake acidification benefits to recreational fishing
Total: all categories		$2,000,000		

Note: See Chapters 5 and 6 of this report for detailed results summaries. Estimates presented with two significant figures; as a result, columns may not add to totals or subtotals.

Table 7-3. Present Value of Monetized Benefits of the CAAA.

	PRESENT VALUE (MILLIONS 2006$)		
	PRIMARY LOW	PRIMARY CENTRAL	PRIMARY HIGH
All Provisions, 1990 to 2020	$1,400,000	$12,000,000	$35,000,000

Note: Values presented in this table are in millions of 2006$, discounted to 1990 using a 5 percent discount rate.

In Table 7-3 we present the mean estimate from the aggregation procedure, along with the Primary Low (i.e., 5th percentile of the distribution) and Primary High (i.e., 95th percentile of the distribution) estimates, for all provisions we assessed. Aggregating the stream of monetized benefits across years involved discounting the stream of monetized benefits estimated for each year to the 1990 present value using a five percent discount rate.

Comparison of Benefits and Costs

Table 7-4 presents summary quantitative results for the prospective assessment, with costs disaggregated by emissions source category and benefits disaggregated by type. We present annual, Primary Central estimate results for each of the three target years of the analysis, with all dollar figures expressed as inflation-adjusted 2006 dollars. The final columns provide net present value estimates for costs and benefits from 1990 to 2020, discounted to 1990 at five percent. The results indicate that the Primary Central estimate of benefits clearly exceeds the costs of the CAAA, for each of the target years and for the cumulative estimates of present value over the 1990 to 2020 period.

As Table 7-4 indicates, a very high percentage of the benefits is attributable to reduced premature mortality associated with reductions in ambient particulate matter and ozone. The CAAA achieves ambient PM reductions through a wide range of provisions controlling emissions of both gaseous precursors of PM that form particles in the atmosphere (sulfur dioxide and nitrogen oxides as well as, to a lesser extent, organic constituents) and directly emitted PM (i.e., dust particles). Because the effects of these constituents on ambient PM are nonlinear, and because some precursor pollutants interact with each other in ways which influence the total concentration of particulates in the atmosphere, separating the effects of individual pollutants on the change in ambient PM would require many iterations of our air quality modeling system. Even with such a tool, the interactive effects of pollutants are complex – as a result the marginal impact of any particular pollutant is dependent on the levels of other pollutants as well. These factors make it difficult to reliably link specific costs to specific aggregate benefits for the pollutant source-specific components of the CAAA (e.g., electric utilities or additional local controls).

Table 7-4. Summary of Quantified Primary Central Estimate Benefit and Costs (Estimates in Million 2006$).

COST OR BENEFIT CATEGORY	ANNUAL ESTIMATES			PRESENT VALUE
	2000	2010	2020	
Costs:				
Electric Utilities	$1,400	$6,600	$10,000	$49,000
Industrial Point Sources	$3,100	$5,200	$5,100	$43,000
Onroad Vehicles and Fuels	$14,000	$26,000	$28,000	$220,000
Nonroad Engines and Fuels	$300	$360	$1,200	$4,500
Area Sources	$660	$690	$770	$7,600
Local Controls to Meet NAAQS	$0	$14,000	$20,000	$53,000
Total Costs	$20,000	$53,000	$65,000	$380,000
Monetized Benefits:				
Avoided Mortality	$720,000	$1,200,000	$1,800,000	$11,000,000
Avoided Morbidity	$27,000	$47,000	$70,000	$410,000
Ecological and Welfare Effects	$15,000	$39,000	$78,000	$310,000
Total Benefits	$770,000	$1,300,000	$2,000,000	$12,000,000

Table 7-5. Summary Comparison of Benefits and Cots.

	ANNUAL ESTIMATES			PRESENT VALUE ESTIMATE
	2000	2010	2020	1990-2020
Monetized Direct Costs (millions 2006$):				
Low[a]				
Central	$20,000	$53,000	$65,000	$380,000
High[a]				
Monetized Direct Benefits (millions 2006$):				
Low[b]	$90,000	$160,000	$250,000	$1,400,000
Central	$770,000	$1,300,000	$2,000,000	$12,000,000
High[b]	$2,300,000	$3,800,000	$5,700,000	$35,000,000
Net Benefits (millions 2006$):				
Low	$70,000	$110,000	$190,000	$1,000,000
Central	$750,000	$1,200,000	$1,900,000	$12,000,000
High	$2,300,000	$3,700,000	$5,600,000	$35,000,000
Benefit/Cost Ratio:				
Low[c]	5/1	3/1	4/1	4/1
Central	39/1	25/1	31/1	32/1
High[c]	115/1	72/1	88/1	92/1
Costs per Premature Mortality Avoided (2006$):				
Central	$180,000	$330,000	$280,000	Not estimated

[a] The cost estimates for this analysis are based on assumptions about future changes in factors such as consumption patterns, input costs, and technological innovation. We recognize that these assumptions introduce significant uncertainty into the cost results; however the degree of uncertainty or bias associated with many of the key factors cannot be reliably quantified. Thus, we are unable to present specific low and high cost estimates.

[b] Low and high benefits estimates based on primary results and correspond to 5th and 95th percentile results from statistical uncertainty analysis, incorporating uncertainties in physical effects and valuation steps of benefits analysis. Other significant sources of uncertainty not reflected include the value of unquantified or unmonetized benefits that are not captured in the primary estimates and uncertainties in emissions and air quality modeling.

[c] The low benefit/cost ratio reflects the ratio of the low benefits estimate to the central costs estimate, while the high ratio reflects the ratio of the high benefits estimate to the central costs estimate.

Table 7-5 provides the results of our more detailed comparison of primary benefits estimates to primary cost estimates. In the top half of the table we show both annual and present value estimates. The cost estimates presented in the table reflect estimates presented in Chapter 3. The monetized benefits indicate both the Primary Central estimate (the mean) from our statistical aggregation procedure and the Primary Low and Primary High estimates (5th and 95th percentile values, respectively). In the bottom half of the table we present three alternative methods for comparing benefits to costs. "Net benefits" reflect estimates of monetized benefits less costs. The table also notes the benefit/cost ratios implied by the benefit ranges, and our estimates of the costs per premature mortality avoided.

The results in Table 7-5 make it abundantly clear that the benefits of the CAAA exceed its costs by a wide margin, making the CAAA a very good investment. Our estimates rely on a particular set of data, models and assumptions we believe are most appropriate at this time.

It is possible that another set of data, models, or assumptions might yield different estimates of benefits, costs, and benefit-cost comparisons. Nonetheless, the very wide margin between estimated benefits and costs, and the results of the uncertainty analysis, suggest that it is extremely unlikely that the monetized benefits of the CAAA over the 1990 to 2020 period reasonably could be less than its costs, under any alternative set of assumptions we can conceive. The central benefits estimate exceeds costs by a factor of more than 30 to one, whether we are looking at annual or present value measures, and the high estimate exceeds costs by roughly 90 to one.

It is also clear from Table 7-5 hat costs fo criteria pollutant programs grow more quickly tha benefits at the beginning of the CAAA compiance period, from 2000 to 2010, and that benefits grow more qickly at the end of the period, from 2010 to 2020. This is consistent with the general statement that ivest ments in clean air tend to inv upfront osts and benefits that accrue over time. The present value estimates in Table show, however, that the total aggregated value of benefits far excees the costs – by our measures, therefore, the programs aociated with the 1990 Clean Air Act Amen dmen have been, and will likely continue to be, a very good investment.

As indicated in the table, the low estimate of net benefits for the year 2020 is positive (i.e., benefits exceed costs) and of significant magnitude - $190 billion. Our uncertainty modeling therefore indicates that the likelihood that the cost estimates of $65 billion in 2020 could exceed the benefits estimates is much less than five percent.

Overview of Uncertainty Analyses

Completion of a study of this breadth and complexity has required EPA to directly confront the role of uncertainty in the key analytic outcomes of the study. While the previous section establishes that the primary estimates of benefits of air pollution control greatly exceed the primary estimates of costs of CAAA compliance, it is nonetheless important to evaluate the extent to which alternative models, assumptions about scenarios, and key parameter choices might affect both benefits and costs. Cognizant of advice to the Agency from the National Research Council, [103] the Project Team developed a three step approach to uncertainty analysis:

1. Identify important sources of uncertainty in each analytical element, starting with emissions profile development. At the end of each of the preceding chapters, we provide a table of key uncertainties and our assessment of the direction and potential magnitude of the impact of this uncertainty on the key analytic output of the study, the monetized net benefits of the CAAA.
2. Quantify parameter and model uncertainty quantitatively where possible by using alternative assumptions or models to estimate intermediate and/or overall net benefit results. In addition, explore options for assessing scenario uncertainty that propagate through the complete analytic chain.
3. Compare the results from these quantitative analyses to the primary results, to inform the degree of confidence in the primary analytic results and to help identify new research directions to address or reduce uncertain and influential components of the analysis.

In the remainder of this section we review each of these three components of our uncertainty analysis.[104]

Identifying Important Sources of Uncertainty

Within each of the summary uncertainty tables in the prior chapters the Project Team has distinguished sources of uncertainty that could have a potentially major impact on the overall net benefits estimate presented in this chapter, based either on quantitative analyses or, where quantitative assessments are unavailable or infeasible, the judgment of Project Team analysts. Potentially major factors are those for which a plausible alternative assumption or approach could influence the overall benefit or cost estimate by five percent or more. We identify a total of 13 potentially major sources of uncertainty in Chapters 2 through 6; these are listed in Table 7-6 below.

Perhaps not surprisingly, the key emissions estimation uncertainties involve forecasting errors, particularly related to estimating future economic and regulatory activity as well as estimating behavior under the counterfactual *without-CAAA* scenario. A key cost estimation uncertainty involves estimating NAAQS compliance, particularly when currently known emissions reductions measures are not sufficient to achieve full compliance with the standard in the future. However, in order for any uncertainty to be considered "major" the impact would need to be of a magnitude of approximately $100 billion to affect net benefits estimates by as much as five percent. In our judgment, while there are several factors that could affect direct cost estimates by a significant percentage, no cost estimation uncertainty has the potential to either more than double our current total cost estimate of $65 billion, or to reduce the cost estimate to $0 or less, which is the magnitude that would be required to constitute five percent of the net benefit estimate.

Several uncertainties that affect benefits estimates, however, could have an impact of $100 billion or greater on the net benefits estimates. Both health effects and valuation uncertainties center on estimation of the impact of air pollutants on mortal risk and the valuation of that health endpoint. The key ecological uncertainty involves identifying what is missing from our necessarily limited quantified ecological benefits. Only one potentially major factor was identified for the air quality modeling step – this may be the result of our inability to apply alternative quantitative air quality modeling tools in this already resource-intensive step in the analytic chain. It is worth noting, however, that as a whole the air quality modeling process very likely contributes a greater than 5 percent uncertainty, of indeterminate direction, to the overall uncertainty in benefits estimates. In addition, the AQMS highlighted uncertainties introduced by the *ex post* adjustment of some primary PM emissions estimates and the procedure used to re-calibrate the CMAQ air quality to account for this emissions adjustment. Although we argue that the overall effect of this source of uncertainty on the net benefits is probably minor (see Table 4-10 in Chapter 4), in some locations ambient PM from primary PM emissions can be more important than secondarily formed fine particles. Overall, we believe that our application of the MATS monitor calibration procedure, which provides a speciated calibration to ensure better agreement between air quality modeling results and comparable monitor data, provides the best agreement possible between our air quality simulation results and monitored values. In the end, however, there is no way to validate the counterfactual, *without-CAAA* scenario estimates.

Table 7-6. Potentially Major Sources of Uncertainty for Estimating the Costs and Benefits of the CAAA.

POTENTIAL SOURCE OF ERROR	ANALYTIC STEP	DIRECTION OF POTENTIAL BIAS FOR NET BENEFITS
Estimated emissions rates under the counterfactual *Without-CAAA* scenario	Emissions	Under-estimate
Estimated economic growth – a key driver of total emissions – under both scenarios	Emissions	Unable to determine
Forecast of the final form and compliance with EPA's revisions of the vacated Clean Air Mercury Rule and the remanded Clean Air Interstate Rule	Emissions	Unable to determine
Secondary organic aerosol (SOA) chemistry	Air Quality Modeling	Under-estimate
Inability to conclusively state that PM mortality outcome is causal based on epidemiology	Health Effects	Over-estimate
Effect of socioeconomic status on mortality from PM exposure	Health Effects	Unable to determine
Attribution of exposure to PM in epidemiology studies based on monitor data	Health Effects	Under-estimate
Omission of short-term effects of PM exposure on mortality	Health Effects	Under-estimate
Timing of reduction in mortality risk after exposure is reduced (cessation lag)	Health Effects	Unable to determine
Source of mortality risk valuation includes many older studies	Valuation	Unable to determine
Scenario of mortal risk in available valuation studies is generally different from that presented by air pollution	Valuation	Unable to determine
Valuation of risk avoidance can change over time and as income increases	Valuation	Unable to determine
Incomplete coverage of ecological effects of air pollutants, including omission of several short-term and virtually all long-term bioaccumulative and persistent effects.	Ecological	Under-estimate

* The classification of each potential source of error reflects the best judgment of the section 812 Project Team. The Project Team assigns a classification of "potentially major" if a plausible alternative assumption or approach could influence the overall monetary benefit estimate by approximately five percent or more. See tables at the end of Chapters 2 through 6 above for more information.

Examination of the last column of Table 7-6 suggests a limited ability to estimate the joint effect of these factors on the direction of potential bias for net benefits. Seven of the factors listed have an indeterminate direction of effect; five yield a potential underestimate of net benefits; and one results in a potential overestimate of net benefits. The large number of factors with an indeterminate direction imply that the direction of the net effect of all factors taken together remains unclear, but the relative confidence that the PM exposure-mortality concentration-response function is causal, based on weight-of-evidence, that being the only uncertainty that yields a potential overestimate, suggests that our primary results may be more likely to understate net benefits than overstate them.

A comparison of the qualitative uncertainty tables from the First and Second Prospective studies indicates that significant advancements over the First Prospective include the use of improved monitoring data for $PM_{2.5}$, an improved understanding and treatment of atmospheric chemistry and the composition of $PM_{2.5}$ emissions, and the use of longer-term simulations with integrated modeling of criteria pollutants using CMAQ rather than a collection of separate air quality models.

Quantifying Model, Parameter, and Scenario Uncertainty

The benefits values presented in this report are subject to a number of uncertainties related to data limitations, analytical choices related to models and input parameters, difficulties predicting future scenarios, and other factors. As noted above, among the most significant model uncertainties is the extensive list of benefits categories, mostly in the ecological area, for which we currently lack the data and/or tools to quantify and monetize benefits. These categories are implicitly treated as having zero value though in reality they may include physical benefits that have a positive economic value. Examples of potentially important, but unquantified ecological effects include nitrogen deposition, non-ozone effects on forest and agriculture vegetation, effects of HAPs on ecological structure and function, and synergistic effects associated with exposures to mixtures of pollutants and interactions of the effects of conventional pollutants such as ozone with climate change. The unquantified and unmonetized benefits thus represent an important underestimation bias in the summary benefit results.

The uncertainties in our quantified and monetized primary benefits estimates that are most likely to significantly influence the primary benefit results are those affecting the largest benefit category: the estimation and valuation of reductions in premature mortality due to decreases in $PM_{2.5}$. Three key uncertainties affecting economic estimates of avoided PM mortality include: (1) the C-R function estimate; (2) the PM/mortality cessation lag structure; and (3) the mortality valuation estimate. These are influential assumptions in our analysis and those for which plausible alternative quantitative estimates are available. The companion Second Prospective Section 812 report, *Uncertainty Analyses to Support the Second Section 812 Benefit-Cost Analysis of the Clean Air Act*, presents detailed quantitative analyses of the sensitivity of benefits results to these and other factors.

Table 7-7 presents a tabular summary of the results of the full range of uncertainty analyses for both costs and benefits, and Figure 7-2 presents a graphical illustration of the impacts of effect of alternative assumptions and models on the central estimate and distribution of monetized avoided mortality benefits, the primary contributor to monetized benefits.

Cost Uncertainties

Table 7-7 shows that the impact of our alternative assumptions about mobile source cost parameters, learning curves, and unidentified local control costs each have relatively modest impacts on total costs, while the I&M failure rate and learning curve assumptions have a slightly larger impact on total costs.[105] In addition, the assumptions underlying our primary cost estimates tend to be conservative; most of the alternatives decrease total compliance costs and none increase costs more than about three percent.

Table 7-7. Quantitative Analyses of Uncertainty in the 812 Second Prospective Analysis.

FACTOR AND LOCATION OF UNCERTAINTY ANALYSIS DISCUSSION IN THIS REPORT (WHERE APPLICABLE)	TYPE OF UNCERTAINTY EVALUATED	ALTERNATIVE ASSUMPTIONS	IMPACT OF ALTERNATIVE ASSUMPTIONS ON 2020 PRIMARY ESTIMATE
UNCERTAINTIES RELATED TO COST ESTIMATES			
Unidentified controls (Chapter 3)	Parameter	Alternate assumption about the threshold for, and cost of, applying unidentified local controls to achieve NAAQS compliance ($10,000/ton).	-18% of local control costs; -2.1% of total costs
I&M program vehicle failure rates (Chapter 3)	Parameter	Alternative assumption about failure rates for I&M program testing based on NRC (2001).	-12% for mobile source costs; -6.5% of total costs
Learning curve assumptions (Chapter 3)	Parameter	Alternate assumptions about the learning rate (5 and 20%)	-6.0% to 3.2% of total costs
Fleet composition and fuel efficiency (Chapter 3)	Scenario	Alternate assumption about future fleet composition and fuel efficiency using AEO 2008.	-3.6% for mobile source costs; -2.0% of total costs
UNCERTAINTIES RELATED TO BENEFITS ESTIMATES			
Alternate C-R function for PM (Chapter 5)	Parameter	Alternative C-R functions – two from empirical liter-ature (Pope et al., 2002 and Laden et al., 2006) and 12 subjective estimates from the expert elicitation study	-83% to 76%, Based on most extreme estimates from PM EE study. Rest of alternatives range from -43% to 41%
Emissions from EGU sources (Chapter 2)	Scenario	Use continuous emissions monitoring (CEM) data in place of Integrated Plan-ning Model (IPM) results, coupled with alternative counterfactual consistent with CEM approach.	+50% in 2000 Due almost entirely to the impact of the alternative *without-CAAA* scenario.
PM/Mortality Cessation lag (Chapter 5)	Model and parameter	Alternative lag structures – one step function and a series of	-23% to 16% when using primary C-R function. -52 to 50%

Table 7-7. (Continued)

FACTOR AND LOCATION OF UNCERTAINTY ANALYSIS DISCUSSION IN THIS REPORT (WHERE APPLICABLE)	TYPE OF UNCERTAINTY EVALUATED	ALTERNATIVE ASSUMPTIONS	IMPACT OF ALTERNATIVE ASSUMPTIONS ON 2020 PRIMARY ESTIMATE
		smooth functions (based on an exponential decay). Smooth functions in some cases also require change in C-R coefficient.	when also changing C-R function.
Value of Statistical Life (Chapter 5)	Parameter	Alternative VSL estimates	-20% to 0%
Discount rates	Parameter	Alternate discount rates (5% and 7%)	-4% to 4%
Alternate C-R function for ozone (Chapter 5)	Parameter	Alternative C-R functions – three from multi-city studies, three from meta-analyses, and one from Jerrett et al. cohort study.	0% for total mortal-ity benefits. -63% to 66% For ozone-related mortality.
Emissions changes by emitting sector	Scenario	Altering each sector-specific emissions by 10 percent	$/ton marginal bene-fit for proportional EGU sector reduc-tions is about 3 times that for nonroad and on-road sectors, and 50% higher than that for area and non-EGU point source sectors.
Differential toxicity of PM components	Parameter	Potential alternative estimates of toxicity for specific PM components	N/A. No quantitative sensitivity analysis performed due to significant data gaps.
Dynamic population modeling (Chapter 5)	Model	Incorporation of dynamic population estimates to calculate life years gained and changes in life expectancy	N/A. Life years gained and changes in life expectancy are Supplemental estimates of PM/mortality effects and cannot be directly compared to the primary estimate.

A further overarching issue with our direct cost methodology is that, for EGU modeling and for some components of the ozone NAAQS compliance cost assessment, the method employed assumes specific optimizing behavior by polluters. In particular, the IPM model used for EGU compliance cost assessment assumes a forward looking approach and may incorporate only limited available information on real-world constraints on optimizing behavior such as long-term fuel supply contracts. If polluters do not optimize in the manner assumed in these models, the direct costs may under-estimate the true costs of compliance. For other emitting sectors, where optimization approaches were not feasible, the potential for under-estimation from this factor does not apply.

A potential issue in considering the uncertainty in cost estimates is our inability to adequately consider the effects of the CAAA on the quality of goods overall. Our method emphasizes that the CAAA does increase the cost of products, and we attempt to hold the quality of products constant in the process. In reality it is likely that the CAAA affects both price and quality of products. One of the more straightforward examples is that motor-vehicle emission controls may reduce performance, though at the same time those controls can increase fuel economy. Other examples include substitution of other devices for charcoal lighting fluid, reformulation of VOC emitting paints, and other product changes that may have altered the quality of those products to consumers. As discussed briefly in Chapter 3, however, the CAAA could also change quality in ways that benefit consumers, but which we do not capture in our estimates – for example, low VOC paint contributes not only to lower ambient ozone levels, but also reduces consumer exposure to VOCs in enclosed indoor environments. Unfortunately, it is very difficult to quantify the effect of this factor on our overall cost and net benefit estimates.

Benefit Uncertainties

On the benefits side, Table 7-7 and Figure 7-2 show that the most influential assumptions affecting benefits are the choice of the C-R function, the cessation lag model for the accrual of benefits, and the VSL distribution. While the two most extreme results from EPA's Expert Elicitation (EE) study imply substantial effects of C-R choice (about 80 percent in either direction) most of the alternatives from the EE study and the published epidemiological studies suggest effects on benefits of about 40 percent or less in either direction. By themselves, longer cessation lag alternatives can reduce monetized benefits by as much as a 23 percent and if coupled with a change in the C-R function, by close to half; however, the Council Health Effects Subcommittee advised that much of the risk reduction benefits from $PM_{2.5}$ controls are more likely to accrue sooner rather than later. Accelerating benefits increases benefits by about 16 percent when maintaining the same C-R function, but could increase them by as much as half when using a smooth function based on the Laden Six Cities follow-up effect estimate. VSL distribution choices in one case produce the same central estimate; in others they reduce VSL between 7 and 20 percent.

A review of the box plots in Figure 7-2 for the factors that have the greatest potential to change the central estimate shows that most of the alternatives do not have a dramatic effect on the spread of uncertainty. Some alternatives suggest the high end of the distribution could be lower, including all of the alternative VSL distributions, which give less weight to higher VSL values than the 26-study Weibull. On the other hand, only a few alternatives (from EPA's particulate matter expert elicitation study) significantly extend the upper end and hardly any extend the lower end, suggesting our primary estimate is unlikely to understate greatly the uncertainty in avoided mortality benefits. In all these cases, however, we are unable to develop a probabilistic representation of uncertainty in the emissions and air quality modeling steps; incorporating uncertainty in these factors would certainly increase the spread between the Primary Low and Primary High estimates.

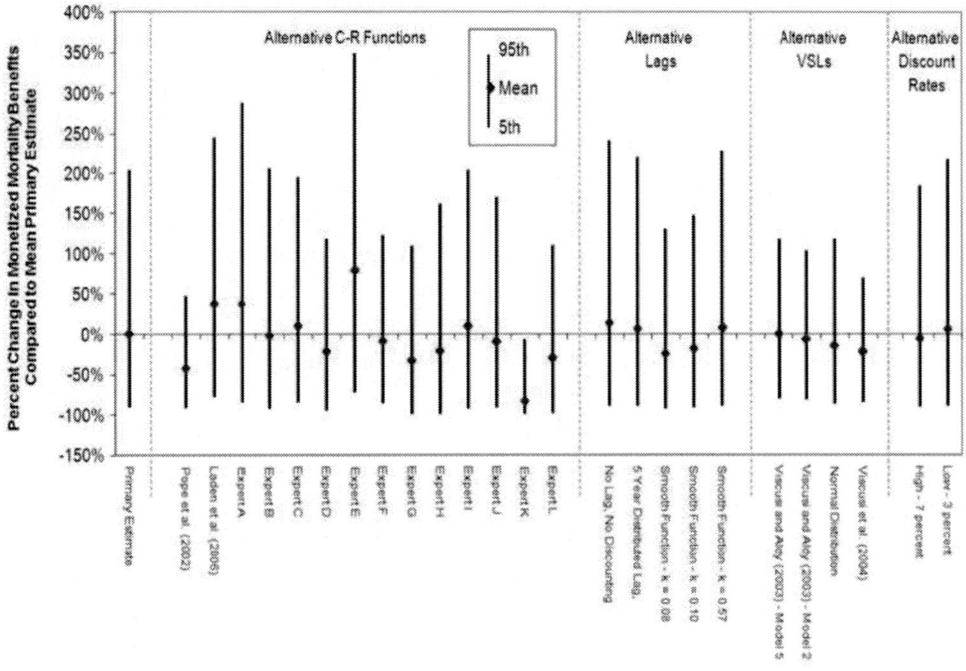

Note: These estimates represent the percent change for each alternative assumption on the total mean mortality benefits estimates. Note that the Project Team assessed the impact Of alternative ozone C-R functions on the total mortality benefits and found that these did not alter the primary estimate. Therefore, alternative ozone C-R functions are not included in this graph.

Figure 7-2. Summary of Quantitative Analysis of Uncertainty in Monetized Mortality Benefits Estimates.

Lessons Learned and New Research Directions

Many of the factors contributing to uncertainty in these estimates are the result of scientific unknowns that might be addressed through additional research. Identification of research directions to address current unknowns can serve an important function - in the First Prospective, for example, we identified eight high priority research directions, six of which were addressed in the Second Prospective.[106]

The key lessons learned in this analysis lead directly to new research directions to inform future assessments, assessments which include both Regulatory Impact Analyses of specific rules and broader, policy-oriented documents such as this Second Prospective. The key insight from this analysis is that rules that target precursors of fine particulate matter are likely to be very cost-effective. Costs per ton of PM control are similar or less than previously estimated, and benefits per ton emitted of these precursors are much larger than previously thought or estimated in the First Prospective. There are three key reasons for the large increase in benefits per ton of PM precursor emitted, involving advances in our understanding of: 1. PM species emissions, 2. the fate of these emissions as estimated by integrated national-scale air quality modeling systems, and 3. the implications of fine particulate air quality for premature mortality.

In addition, the results of the study also provide evidence of the significant benefits of avoiding mortality associated with ozone exposure, avoiding degradation of visibility in residential and recreational settings, and avoiding significant chronic and acute morbidity, including chronic bronchitis and acute myocardial infarction. The last two of these monetized benefits categories were shown, by themselves, nearly to equal the full costs of all provisions of the CAAA. There also remain large categories of health and ecological benefits for which we have no quantified or monetized benefit estimates. For example, although there is an established literature linking air pollutant exposure with increased risk of cerebrovascular accidents (stroke), as well as a literature on the medical costs of this condition, that category of effect is not yet included in our estimates of the health benefits of reducing air pollutant exposure.

Insights gleaned from completing this study suggest the following eight areas to be the highest priority research needs:

- ***Improving cost analyses for rules that are technology-forcing***. The overall cost analysis in Chapter 3 is characterized by complete coverage of the costs of many rules, but the Project Team acknowledges that in some cases, particularly involving compliance with tighter future NAAQS standards, application of the suite of known, cost-effective current pollutant control measures are not sufficient to achieve compliance in all locations. This shortcoming remains one of the important focal points for compliance cost research within the Agency. One possible direction that the Agency is considering is analysis of historical data on the cost and penetration rates of new emissions control technologies, particularly those for NAAQS compliance, which could provide insights on the process, cost, timing, and potential limits of induced innovation.
- ***Continuing efforts to incorporate a broader range of market benefits in economy-wide modeling of the impacts of regulation***. The results of Chapter 8 indicate that there are significant benefits to economic growth when we consider the labor force and health expenditure implications of cleaner air. Our demonstration of the importance of incorporating benefits-side effects in macroeconomic modeling efforts, however, does not incorporate all possible market effects of cleaner air. For example, increased agricultural and forest productivity might feasibly be incorporated in the model we employed. Ultimately, it will also be important to develop new methods to characterize the large nonmarket benefits of cleaner air in these models, including most importantly the welfare enhancements (as opposed to simply the market implications) associated with reductions in premature mortality.
- ***Understanding synergies and antagonistic effects of climate change in realizing benefits, as well as for understanding co-benefits of greenhouse gas (GHG) control policies***. Consideration of climate change was outside the scope of this Second Prospective effort, but designing effective and efficient regulatory mechanisms for GHG emissions control has rapidly become an important priority for the Agency. The methods, data, and results of this study are important for modeling co-benefits of GHG control policies, as many policies targeted at GHG reductions also reduce other, conventional pollutants, and those benefits are realized sooner than the generally long-term benefits of GHG policies. In addition, climate change likely alters the benefits achieved by conventional pollutant policies, as for example

increases in mean temperature as well as increases the frequency of extreme temperature events creates conditions conducive to ozone formation. Both areas are important for further research.

- ***Developing probabilistic representation of emissions and air quality to support uncertainty analysis***. As noted earlier in this chapter, a major shortcoming of existing quantitative characterizations of uncertainty in benefits and costs of the CAAA is the inability to integrate uncertainties in emissions and air quality modeling steps. Two areas of research deserve further attention: 1. Developing more nimble tools for assessing the air quality implications of emissions control policies, or updating those that exist; 2. Developing probabilistic characterizations of key parameters that contribute to overall uncertainty in emissions and air quality analyses. Pursuit of the latter initiative will likely require application of expert elicitation, either formal or informal, to make progress.
- ***Understanding the potential for differential toxicity to play a role in benefits of control programs and, by extension, policy priorities.*** The issue of species-specific particulate matter toxicity remains very complex, involving the effects of mixtures and synergies of species that are not currently well understood. It is nonetheless important to understand the extent to which rules targeted at specific PM species might yield similar benefits as rules targeting total PM mass.
- ***Continuing to pursue evidence of the real-world public health impact of specific air quality actions.*** Sometimes referred to as *accountability analyses*, tracking the real-world instances of rapid air quality changes, either improvements or reductions in air quality, can yield important corroborating evidence of the effects found in epidemiology studies. As we found in our uncertainty analyses supporting the Second Prospective, these natural experiments also provide insights for the nature of cessation lags, and might be useful in better understanding species-specific toxicity.
- ***Expanding coverage of ecological benefits.*** There are potentially large ecological benefits of air pollution control that are not currently quantified. Some of the most important categories of unquantified effects include nitrogen deposition effects on estuarine areas, sulfur deposition effects on vegetation and other aspects of terrestrial systems, and long-term effects of air toxics. Perhaps equally important, but much more subtle, are the long-term effects of a wide range of air pollutants on ecosystem structure and function. Even potentially beneficial effects of pollutants, such as deposition of the nutrient nitrogen in terrestrial and even actively managed farms and forests, might have longer-term detrimental effects on nutrient cycling and species selection that are currently poorly understood.
- ***Expanding coverage of health benefits.*** Great effort has been expended to better characterize the full range of health implications of air pollution. Despite this effort, it is still difficult to quantify the link between air pollution and stroke, and it is also difficult to assess the incremental effects of gaseous pollutant exposures, in part because there are only a limited set of studies that characterize the individual contributions of multiple pollutant exposures on health outcomes. While the Agency has developed robust benefits analyses for programs that control individual gaseous pollutants, such as carbon monoxide, it remains difficult to incorporate these effects in multi-pollutant models that include PM, ozone, and other gaseous pollutants typically present in many settings in the U.S.

The results of this Second Prospective study clearly provide strong evidence that the nation's investment in clean air has been a wise and cost-effective policy. Continued effort is needed to ensure that air pollution policies are pursued in the most cost-effective manner possible. Pursuit of these research goals should continue to enhance our ability to provide accurate and timely assessments of the costs and benefits of all provision authorized under the Clean Air Act and its Amendments.

8. COMPUTABLE GENERAL EQUILIBRIUM ANALYSIS

The 1990 Clean Air Act Amendments (CAAA) represent a significant change in Federal air pollution policy affecting virtually every sector of the U.S. economy, including industry as well as individual households. The cost and benefit estimates presented in the previous chapters reflect the direct impacts of the CAAA in terms of industry's and households' direct compliance expenditures and the value of the direct human health, visibility, ecological, and other benefits associated with CAAA-related improvements in air quality. The cost-benefit information is central to EPA's analysis of the Amendments, but policymakers and the public are also interested in the impact of CAAA programs on overall economic performance. Therefore, to supplement the direct cost and benefit estimates presented in the previous chapter, the Project Team applied an economy-wide computable general equilibrium (CGE) analysis of the Amendments and estimated the effect of the CAAA on U.S. gross domestic product and other macroeconomic measures. The Project Team performed this analysis with the Economic Model for Policy Analysis (EMPAX-CGE), a CGE model employed by EPA for several previous analyses of CAAA regulations, including the National Ambient Air Quality Standards (NAAQS) for $PM_{2.5}$, the 8-hour Ozone NAAQS, and the Clean Air Interstate Rule.

The Project Team's CGE analysis for the Second Prospective represents a major step forward in EPA's application of CGE models in the context of air pollution policy. Unlike previous CGE analyses that focused exclusively on the macroeconomic impacts of compliance expenditures, the Second Prospective incorporates impacts related to both CAAA costs and some categories of benefits into EMPAX-CGE, to the extent feasible. Because both the costs and benefits of CAAA regulations may affect the size and composition of the U.S. economy, the Project Team's approach provides a more comprehensive and balanced view of the macroeconomic impacts of air pollution policy than previous assessments. To illustrate the extent to which including labor force and medical expenditure impacts in EMPAX-CGE affects model results, we applied the model in two ways: one model run that reflects only the costs of the CAAA (the cost-only case) and a second model run that reflects both the costs and a subset of the total benefits of the Amendments (the labor force-adjusted case).

This chapter presents the CGE analysis in four sections. In the first section, we provide an overview of EMPAX-CGE, describing the model's overall structure and highlighting the sectoral and geographic resolution of the model. The second section describes the development of the cost- and labor force and health expenditure benefit-side inputs for the analysis and documents how these inputs were incorporated into EMPAX-CGE. The third section presents the results of our analysis, both in aggregate and by industry. To conclude the chapter, we discuss the major uncertainties of the analysis and their implications for results.

EMPAX-CGE[107]

EMPAX-CGE is a multi-industry, multi-region computable general equilibrium model of the U.S. economy. Below we describe the main features of the typical CGE model, followed by a more detailed overview of the structure and functionality of EMPAX-CGE.

Overview of CGE Modeling

CGE models simulate the flow of commodities and factors of production (i.e., labor, capital, and natural resources) among producers and households to assess how a change in policy or an economic shock affects the size and composition of the economy. As shown in Figure 8-1, households in CGE models own factors of production (capital, labor, and natural resources) that they supply to firms in exchange for wages and other forms of income. Firms use these factors in conjunction with intermediate inputs purchased from other industries to produce goods and services, which are sold to other industries as well as consumers. Goods and services can also be exported, and imported goods can be purchased from other countries.

In modeling the circular flow of the economy depicted in Figure 8-1, CGE models capture behavioral changes among households and firms in response to changes in prices. At the producer level, CGE models simulate the substitution of inputs as the price of one input, such as steel or labor, rises relative to the price of other inputs. This allows the simulation of producer behavior in CGE models using minimization of the cost of production as an objective, consistent with the behavior of firms in the real economy. Similarly, as the price of one good rises relative to the prices of other products and services, CGEs model the process whereby households consume less of the more expensive good and more of other goods. Related to households' substitution between different goods, CGE models also simulate household substitution between labor and leisure as real wages change. Because the productive capacity of the economy is dependent, in part, on labor supply, the labor-leisure tradeoff is critical in determining the size of the economy.

The general equilibrium component of CGE modeling requires a comprehensive market coverage in which all sectors in the economy are in balance and all economic flows are accounted for. Establishing equilibrium conditions requires that every commodity that is produced must be purchased by firms or consumers within the United States or exported to foreign consumers. The requirement for all markets to be in equilibrium during the time period of the model simulation is a simplifying assumption of the model, but is nonetheless a condition which, over time, is consistent with production in the actual economy. Prices of these goods reflect all costs of production. Households receive payments for their productive factors and transfers from the government (not shown in Figure 8-1), and this income must equal consumer expenditures and savings. In aggregate, all markets must clear, meaning that supplies of commodities and factors must equal demand, and the income of each household must equal its factor endowments plus any net transfers received. An important implication of this market clearing assumption is that CGE models assume that the economy is at full employment (i.e., there is no involuntary unemployment). Therefore, CGE models do not typically provide insights into the unemployment impacts of policy changes.

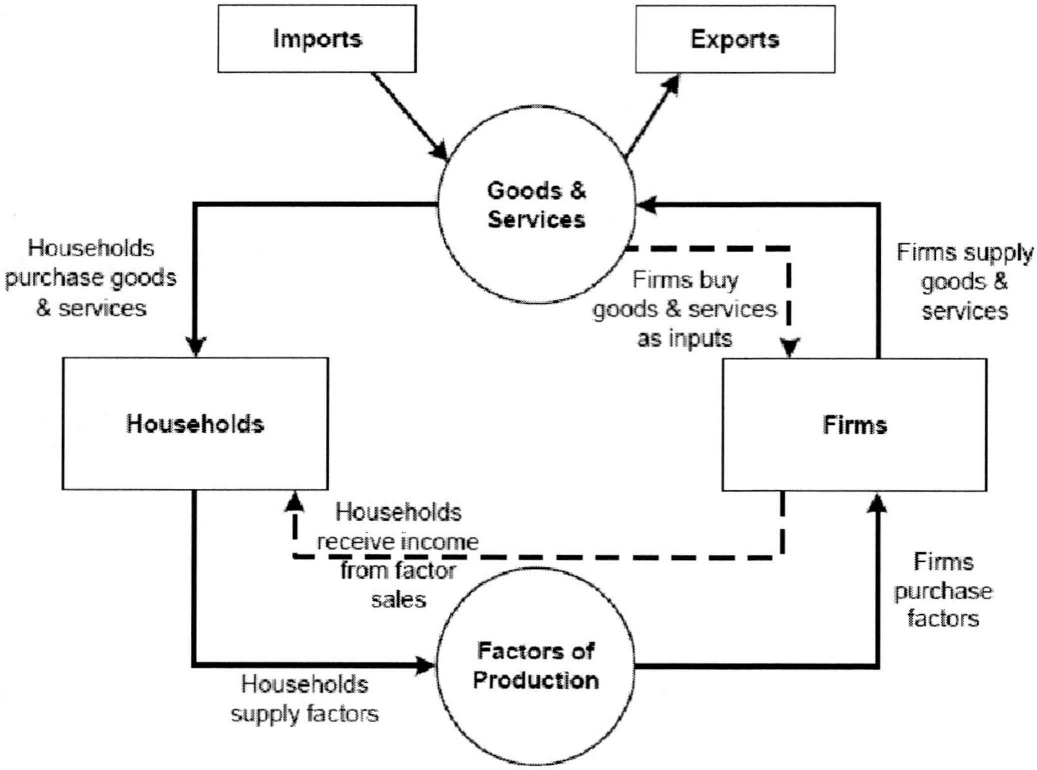

Source: RTI International, EMPAX-CGE Model Documentation, prepared for U.S. EPA Office of Air Quality Planning and Standards, March 2008.

Figure 8-1. CGE Model Schematic.

Overview of Empax-CGE

Similar to other CGE models, EMPAX-CGE is structured to represent the complex interactions between consumers and producers in the real economy. To model these interactions, EMPAX-CGE performs thousands of calculations with the objective of maximizing household utility (well-being) while simultaneously maximizing firm profits. While complex, these calculations are a simplified representation of the real economy. The behavior of households and firms is inherently multi-faceted and dependent on a range of factors, many of which are not well understood. To model this behavior, EMPAX-CGE uses a simplified, hierarchical representation of household and firm decision-making that reduces the behavior of households and firms to a limited number of structured decisions. For example, as shown in Figure 8-2, the first decision for the household sector in EMPAX-CGE is the optimization of consumption and leisure. To model this decision, EMPAX-CGE assumes that households are free to allocate their time between labor and leisure to maximize their welfare. Time that households do not devote to leisure represents household labor supplied to producers. Therefore, in effect, the leisure-consumption decision also represents a tradeoff between leisure and labor force participation. After the consumption-leisure decision, EMPAX-CGE simulates household consumption as a series of hierarchical decisions involving consumption goods and transportation.

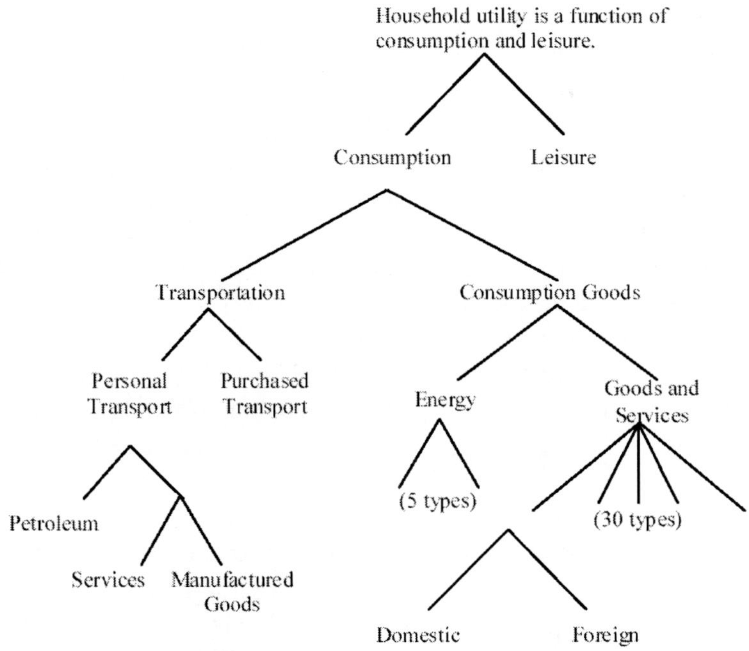

Figure 8-2. Empax-CGE Decision Hierarchy for Households.

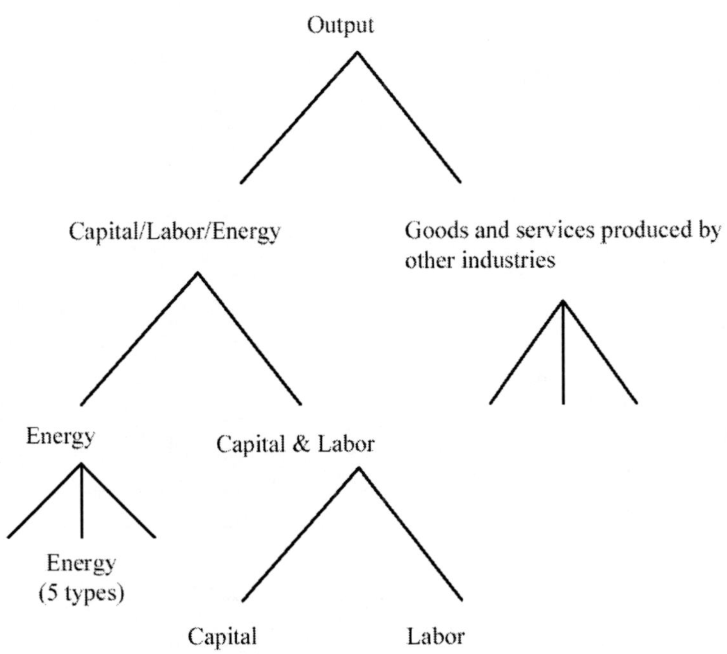

Figure 8-3. Empax-CGE Nested Structure for Producers.

EMPAX-CGE also models *firm* behavior as a series of hierarchical decisions. Similar to EMPAX-CGE's treatment of households, this hierarchical structure represents a simplification of how firms decide which inputs to use in the production of goods and services. As illustrated in Figure 8-3, the first tier of this decision hierarchy is a choice between: (1) an indeterminate mix of capital, labor, and energy and (2) goods and services produced by other industries, such as steel or computer equipment. Producers then optimize among capital, labor, and energy.

Consistent with simplifying household and firm decision-making into the structured frameworks depicted in Figures 8-2 and 8-3, EMPAX also uses a simplified representation of the overall structure of the economy. Firms in the U.S. are scattered across thousands of industries and produce countless goods and services. Modeling each of these sectors individually within an economy-wide model, however, is not feasible due to data and computational processing constraints. To address this issue, EMPAX-CGE aggregates the economy into 35 distinct industries, as listed in Table 8-1. The industry classifications included in EMPAX-CGE were defined so as to maximize the level of sectoral detail among energy-intensive and manufacturing industries. EMPAX-CGE also separates the electricity industry into fossil fuel generation and non-fossil generation, which is important for assessing the impacts of policies that affect only fossil fuel-fired electricity, such as air pollutant regulations.

Table 8-1. Industries in Empax-CGE.

EMPAX Industry	North American Industry Classification System (NAICS)
Energy	
Coal	2121
Crude oil[a]	211111, 4861
Electricity (*fossil and nonfossil*)	2211
Natural gas	211112, 2212, 4862
Petroleum refining[b]	324, 48691
General	
Agriculture	11
Mining (w/o coal, crude, gas)	21
Construction	23
Manufacturing	
Food products	311
Textiles and apparel	313, 314, 315, 316
Lumber	321
Paper and allied	322
Printing	323
Chemicals	325
Plastic and rubber	326
Glass	3272
Cement	3273
Other minerals	3271, 3274, 3279
Iron and steel	3311, 3312
Aluminum	3313
Other primary metals	3314, 3316
Fabricated metal products	332

Table 8-1. (Continued)

EMPAX Industry	North American Industry Classification System (NAICS)
Manufacturing equipment	333
Computers & communication equipment	334
Electronic equipment	335
Transportation equipment	336
Miscellaneous remaining	312, 337, 339
Services	
Wholesale & retail trade	42, 44, 45
Transportation [c]	481-488
Information	51
Finance and real estate	52, 54
Business/professional	53, 55, 56
Education (w/public)	61
Health care (w/public)	62
Other services	71, 72, 81, 92

[a] Although NAICS 211111 covers both crude oil and gas extraction, the gas component of this sector is addressed in the natural gas energy sector.

[b] EMPAX-CGE reports output for the petroleum refining industry based on the delivered price of petroleum products. This reflects the value of pipeline transport.

[c] Transportation does not include NAICS 4862 (natural gas distribution), which is part of the natural gas industry.

EMPAX-CGE is also designed to reflect regional differences in the overall structure of the economy. Because the availability and cost of different production inputs, such as labor and energy, vary across different regions of the U.S., the response of a given industry to changes in policy may vary by region. To account for this effect, EMPAX-CGE models each industry separately in five different regions, as shown in Figure 8-4. The specification of the five economic regions included in the model is based, as closely as possible, on the structure of the electricity market regions defined by the North American Electric Reliability Council (NERC).[108]

EMPAX-CGE assumes that households have perfect foresight of future changes in policy and maximize utility over the full time horizon of the model. To adjust to future policy changes, households may alter their decisions about labor force participation and modify their consumption patterns in terms of their overall level of consumption and the mix of goods and services they choose to consume. This is in contrast to static CGEs, which model the economy without regard for time (i.e., they effectively model the economy for a single time period).

EMPAX-CGE contains four representative households in each model region, classified by income. These household income groups are:

- $0 to $14,999,
- $15,000 to $29,999,
- $30,000 to $49,999, and
- $50,000 and above.

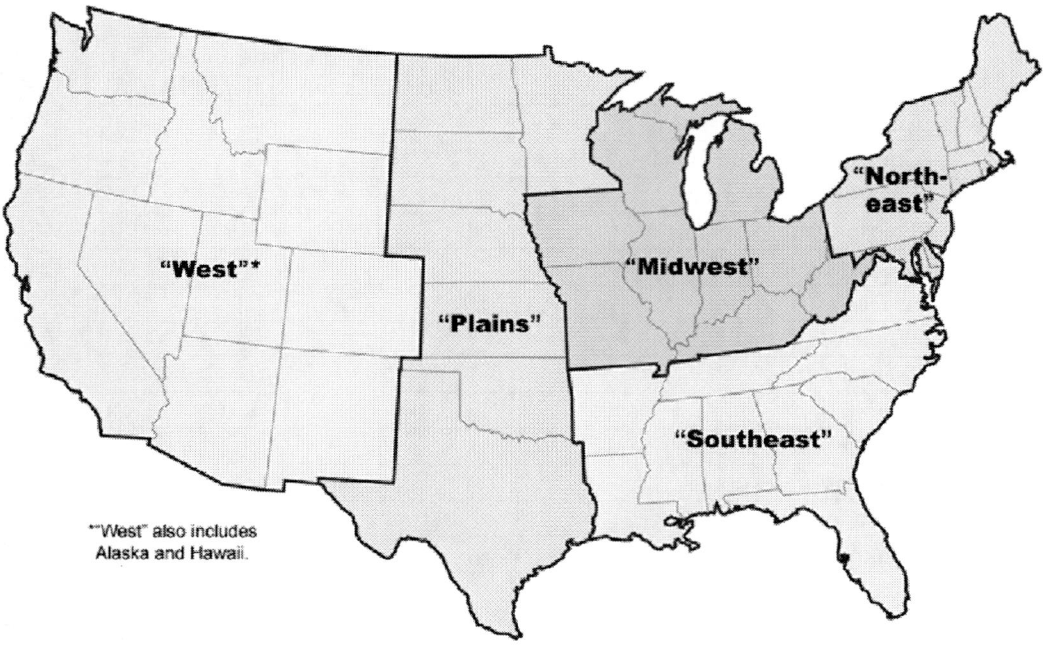

Figure 8-4. Empax-CGE Regions.

These representative households are assumed to possess certain factors of production including labor, capital, natural resources, and land inputs to agricultural production. Factor prices are equal to the marginal revenue received by firms from employing an additional unit of labor or capital, and households allocate income from sales of these productive factors to purchases of consumption goods to maximize welfare.

The outputs generated by EMPAX-CGE include GDP, consumption, and an economic welfare measure known as Hicksian equivalent variation (EV). EV is based on the concept of willingness-to-pay, which is the maximum amount a household would pay for a particular good or service (including leisure), given its budget constraint. Willingness to pay reflects the value or welfare that a household derives from the consumption of a good or service. For a given policy scenario, the change in EV represents the additional money that a household would require (at original prices and income) to make it as well off with the new policy as it was under baseline conditions; this amount is "equivalent" to the change in utility the household derives from consumption and leisure time. It is important to note, however, that EMPAX-CGE's estimation of EV captures welfare associated with market goods and services but does not capture non-market effects. As a result, the measure would not reflect some categories of household welfare that are important to our cost-benefit analysis, such as avoided pain and suffering associated with health effects incidence, improvements in visibility, and changes in service flows that derive from well functioning ecological resources.

The baseline values for the outputs generated by EMPAX-CGE are adapted from the economic forecast in the Department of Energy's *Annual Energy Outlook 2007*. These baseline values represent the U.S. economy under the *with-CAAA* scenario for the Second Prospective.[109]

Development of Model Inputs

The Project Team estimated the macroeconomic impacts of the CAAA as the difference between (1) the EMPAX-CGE reference case projections, which represent the *with-CAAA* scenario, and (2) EMPAX-CGE projections for the *without-CAAA* scenario. To conduct the model runs for the *without-CAAA* scenario, the Project Team developed model inputs related to both the costs and benefits of the Amendments. To assess the difference in costs associated with CAAA compliance, we estimated CAAA-related compliance expenditures by industry and EMPAX region. Based on these estimates, the Project Team reduced the cost of production for affected industries from the baseline costs of production to develop industry-wide cost structures for the *without-CAAA* scenario. The "cost-only" runs therefore estimate the loss in economic productivity associated with CAAA compliance costs.

As noted above, however, the CAAA also yields benefits that result in potentially substantial changes in economic production as well. The benefit-side inputs developed by the Project Team include (1) medical expenditures associated with pollution-related illness, (2) the change in workers' time endowment due to pollution-related mortality, and (3) the change in workers' time endowment due to pollution-related morbidity. The Project Team incorporated changes in medical expenditures into EMPAX-CGE as changes in household expenditure patterns. To incorporate changes in the amount of time workers can devote to labor or to leisure in the model, we first estimated how health effects and mortality estimated in Chapter 5 would affect the exposed population's ability to supply labor to firms. Estimates of lost work time associated with morbidity have been estimated in prior work or are available from BenMAP.[110] Next, we assumed that pollution-related illness and mortality among the labor force reduce workers' overall time endowment (labor and leisure) in proportion to the effect on labor supply. That is, if air pollution would reduce labor supply by x percent in 2020, the Project Team assumed that the overall time endowment of workers would also decline by x percent in 2020.

We did not attempt to incorporate time endowment effects for people outside the formal economy (e.g., retirees, students, homemakers) into EMPAX-CGE. While the "nonworking" population is clearly affected by air pollution, and those effects are likely to influence the level and composition of economic activity, the structure of EMPAX-CGE is not conducive to assessing how these populations affect the economy. The results presented in this chapter therefore likely underestimate the macroeconomic impacts resulting from CAAA-related improvements in public health.

Below we describe the Project Team's approach for generating the EMPAX-CGE inputs related to the costs and benefits of the CAAA. As noted above, the Project Team used these inputs to conduct two analyses of CAAA-related macroeconomic impacts; the first reflects only the costs of the CAAA (the cost-only case), while the second reflects both the costs and selected human health benefits of the Amendments (the labor force-adjusted case).

Cost Inputs

To assess the macroeconomic impacts of CAAA-related costs, the Project Team incorporated CAAA compliance expenditures by industry and region into EMPAX-CGE. Similar to other CGE models, EMPAX-CGE is an expenditure-based model and therefore requires expenditure-based inputs to represent the costs of the Amendments. CAAA compliance expenditures, however, are not always the equivalent of the direct costs of the

Amendments presented in Chapter 3. While the direct costs of the CAAA reflect the value of the capital, labor, and other resources necessary for CAAA compliance, compliance expenditures simply represent the financial resources exchanged for CAAA compliance. For example, the direct costs of the Amendments do not include taxes, because such payments represent transfers rather than resources expended to control air pollutant emissions. In contrast, CAAA compliance expenditures *include* transfers because they represent an exchange of financial resources from one party (e.g., a firm) to another (e.g., the government) that can affect the choices made by firms.

To estimate the compliance expenditures associated with the Amendments, the Project Team made three adjustments to the direct cost estimates presented in Chapter 3:

1. ***Inclusion of fuel excise taxes:*** The Project Team included fuel excise taxes in the compliance expenditure estimates developed for the EMPAX-CGE analysis. Excise taxes were excluded from the direct cost estimates presented in Chapter 3 because such taxes are transfers.
2. ***Industry-specific discount rates:*** Unlike the direct cost estimates presented in Chapter 3, which reflect a 5 percent social discount rate, the compliance expenditures presented in this chapter reflect the private discount rates of affected industries. For each industry, we estimated the private discount rate based on the industry-specific weighted average cost of capital as reported in Ibbotson Associates' *Cost of Capital Yearbook.*[111]
3. ***Exclusion of motorist waiting time from cost estimates for inspection and maintenance programs:*** The direct cost estimates for motor vehicle inspection and maintenance (I&M) programs in Chapter 3 reflect the value of motorist waiting time. Although waiting time represents a welfare loss to society, this cost is not incurred as an expenditure. Because CGEs are expenditure-based models, we exclude motorist waiting time from the cost-side inputs incorporated into EMPAX-CGE. The exclusion of motorist waiting time is unlikely to significantly affect the results of the CGE analysis, as these costs represent only 18 percent of direct CAAA costs associated with I&M programs and less than 5 percent of direct costs for the entire on-road sector.

Based on these adjustments, we developed the compliance expenditure estimates presented in Table 8-2. For comparison, the exhibit also includes the direct cost estimates summarized in Chapter 3. As indicated in the exhibit, the estimated CAAA compliance expenditures in 2010 are approximately $2.0 billion greater than the Project Team's direct cost estimates for 2010. In 2020, the difference between the two increases to $3.0 billion. The estimates in Exhibit 8-6 also show that the distribution of compliance expenditures across source categories is similar to the distribution of direct costs.[112]

Benefit Inputs

As noted above, the Project Team's analysis of the macroeconomic impacts of CAAA-related health improvements focuses on three specific effects: (1) the change in the household time endowment from pollution-related mortality impacts, (2) the change in the household time endowment from pollution-related morbidity, and (3) the change in medical expenditures associated with pollution-related morbidity. The Project Team incorporated these effects into the without-CAAA EMPAX-CGE model runs to estimate the size and composition of the economy in the absence of the Amendments. The methods employed to quantify these effects and convert them into useable inputs for EMPAX-CGE are described below.

Mortality-Related Labor Force Impacts

The Project Team incorporated pollution-related mortality impacts into EMPAX-CGE as a percentage change in the time available to workers for labor and leisure activities (i.e., their time endowment). In estimating this percentage change, the Project Team focused on the dynamic population effects of premature mortality from particulate matter (PM) exposure. While ozone also leads to premature mortality, the benefits results in Chapter 5 show that reductions in ambient PM concentrations are responsible for approximately 98 percent of the avoided cases of premature mortality associated with the Amendments in both 2010 and 2020. Because of the dominant effect of PM on mortality (relative to ozone) and the lack of tools available to examine the dynamic population effects of PM and ozone in an integrated fashion, the Project Team focused the mortality component of the EMPAX-CGE analysis on changes in PM-related mortality.

The mortality-related inputs developed by the Project Team reflect the dynamic effects of PM mortality on the population over time. When PM concentrations change, the resulting population impact grows over time, as the change in population for any given year reflects changes in the incidence of PM-related mortality from prior years. For example, if PM concentrations are reduced permanently in 2015, the population (and the size of the labor force) in 2017 will reflect avoided cases of premature mortality in 2015, 2016, and 2017. Over time, this dynamic effect leads to a significant number of life years saved as the reduction in pollution-related risk is applied to successively larger populations each year (due to previous years' improvements in air quality).

Table 8-2. Summary of Annual CAAA Compliance Expenditures and Direct Costs (Millions of 2006$).

	2010		2020	
SOURCE CATEGORY	COMPLIANCE EXPENDITURES (USED FOREMPAX ANALYSIS)	DIRECT COSTS	COMPLIANCE EXPENDITURES (USED FOR EMPAX ANALYSIS)	DIRECT COSTS
Electric Generating Units	$8,470	$6,640	$13,000	$10,400
On-road Sources	$24,800	$25,800	$27,200	$28,300
Non-road Sources	$750	$359	$1,620	$1,150
Industrial Point Sources	$5,580	$5,180	$5,600	$5,140
Area Sources	$693	$693	$768	$767
Local Controls (Identified)	$5,590	$5,250	$6,790	$6,180
Unidentified Local Controls	$9,020	$9,020	$13,500	$13,500
TOTAL	**$54,900**	**$52,900**	**$68,500**	**$65,500**

To capture these dynamic effects, the Project Team used a spreadsheet-based dynamic population simulation model described in Chapter 5.[113] The model was designed to track the effect of alternative assumptions about the mortality effects of $PM_{2.5}$ on the U.S. population, but may also be used to assess how changes in $PM_{2.5}$ concentrations lead to changes in the population over time. The tool incorporates detailed life table data for historical years, by age, gender, and cause of death, obtained from the Census Bureau and the Centers for Disease Control. It also incorporates Census mortality and population projections for future years, again by age and gender, using the projected death and birth rates that underlie the Census Bureau's published population projections. For a given model scenario, the model simulates the U.S. population by single year age group and gender for each year through 2050.

To estimate changes in the labor force with the population simulation model, the Project Team employed the following three-step approach:

1. ***CAAA-related change in population:*** First, the Project Team entered changes in $PM_{2.5}$ concentrations into the population simulation model based on the air quality modeling analysis described in Chapter 4. Netting the model results from baseline (*with-CAAA*) population projections, the Project Team estimated PM-related changes in population by gender and single-year age group for both the 2010 and 2020 target years (and for every other year in the model time horizon). These changes represent the estimated difference in population between the *with-CAAA* and *without-CAAA* scenarios.
2. ***CAAA-related change in the labor force:*** To estimate the change in the labor force associated with the CAAA, the Project Team applied age- and gender-specific labor force participation rates from the Bureau of Labor Statistics to the changes in population estimated in Step 1.
3. ***Percent Change in Labor Force:*** The Project Team estimated the percent change in the labor force associated with pollution-related mortality by dividing the total labor force changes estimated in Step 2 by baseline (with-CAAA) projections of the total labor force. As indicated above, the Project Team assumes that this percent change applies to the full time endowment (labor and leisure time) for the labor force.

Morbidity-related Labor Force Impacts

Similar to pollution-related mortality, pollution-related morbidity was incorporated into EMPAX-CGE as a percent change in the labor and leisure time available to workers. Unlike the Project Team's PM-based approach for mortality, the approach for morbidity accounts for both PM- and ozone-related impacts. The literature for the various PM and ozone endpoints examined use several different metrics for quantifying labor force impacts. To standardize these estimates, we converted the values obtained from the literature to the number of work days lost per case, by endpoint. We then applied these values to the yearly changes in the number of cases for each endpoint to estimate the total work days lost for any given year. These values reflect the labor force participation rate among those individuals afflicted by each health effect. Because the time endowment in EMPAX-CGE measures time on an annual basis, we converted the estimated number of work days lost to lost work years, based on an assumed work year of 235 work days.[114] To express work years lost as a percent change in the labor force, we divided the estimated work years lost for each target year by the projected size of the labor force. The resulting value represents the percent change in workers' labor time.

Table 8-3. Work Days Lost Per Case, by Morbidity Endpoint[1].

PM[2]		
Acute Myocardial Infarction[3]	Age <25: N/A Age 25-34: 17.7 days Age 35-44: 14.5 days	Age 45-54: 23.7 days Age 55-65: 137.0 days Age>65: 0 days
Chronic Bronchitis[3]	Age <25: N/A Age 25-34: 50.3 days Age 35-44: 42.2 days	Age 45-54: 55.5 days Age 55-65: 73.5 days Age >65: 0 days
Hospital Admissions, Cardiovascular[4]	Age 0-14: N/A Age 15-44: 18.3 days	Age 45-64: 17.9 days Age >64: 7.0 days
Hospital Admissions, Respiratory[4]	Age 0-14: N/A Age 15-44: 30.7 days	Age 45-64: 30.1 days Age >64: 7.5 days
Emergency Room Visits, Respiratory[5]	Average across all age groups: 0.2 days	
Work Loss Days	Average among working age population: 1 day	
Ozone[6]		
School Loss Days[7]	Average across all age groups: 0.7 days	
Worker Productivity	Not applicable[8]	
Hospital Admissions, Respiratory[9,10]	Age <2: 0 days Age >64: 7.5 days	
Emergency Room Visits, Respiratory[5]	Average across all age groups: 0.2 days	

Notes: N/A indicates that the underlying C-R function does not provide incidence estimates for that age group.

1. Except for chronic bronchitis and acute myocardial infarction, the number of work days lost is applied to the change in annual incidence. For chronic bronchitis and acute myocardial infarction, the work days lost presented in this table are applied to annual changes in the prevalence of each disease.
2. We did not generate separate work loss day estimates for the following PM health endpoints discussed in Chapter 5: acute bronchitis, acute respiratory symptoms, asthma exacerbation, lower respiratory symptoms, and upper respiratory symptoms. The lost work days associated with these endpoints are already reflected in the work loss day endpoint included in this table.
3. Derived from Cropper and Krupnick (1999).
4. Agency for Healthcare Research and Quality (2000), as cited in BenMAP user's guide, Abt Associates (2008).
5. We assume that each E.R. visit equals one day of lost work time per worker affected. The estimate of 0.2 days per case reflects the percentage of cases realized by the working-age population, the ratio of workdays to total days in a year (235/365), and the percent of the working-age population in the labor force.
6. We did not estimate the number of work days lost per case of acute respiratory symptoms associated with ozone exposure.
7. Derived from Abt Associates (2008). Note that 0.7 is the estimated average work loss days per school loss day, incorporating work-force participation rates for caregivers.
8. The benefits analysis presented in Chapter 5 does not estimate the number of cases for the worker productivity endpoint. Instead, worker productivity is estimated as the change in income associated with changes in ozone concentrations. We estimated the work days lost per dollar of income lost based on the average daily wages of outdoor workers.
9. Derived from Abt Associates (2008).
10. The dose-response function for ozone-related respiratory hospital admissions does not cover populations older than two years old and younger than 65. For this endpoint we do not address potential caregiver time lost for incidence in either cohort.

As suggested above, estimating the number of work days lost per case for each endpoint is a key step in the Project Team's methodology. Table 8-3 summarizes these endpoint values for both PM and ozone. With the exception of chronic bronchitis and acute myocardial infarction (AMI), the estimates presented in Table 8-3 were applied to the annual change in incidence for each endpoint (i.e., the change in the number of new cases per year), as the duration of disease for most endpoints is no more than several weeks. Chronic bronchitis and AMI, however, affect individuals over multi-year time horizons. We therefore apply the work loss day estimates for these endpoints to changes in the prevalence of each disease (i.e., the change in the number of people with the disease, relative to the baseline).

Table 8-4. Annual Medical Expenditures Per Case, by Morbidity Endpoint (2006$).[1]

	2010	2020
PM[2]		
Acute Myocardial Infarction[3]	$17,600	$17,300
Chronic Bronchitis[4]	$715	$810
Emergency Room Visits, Respiratory[5]	$369	
Hospital Admissions, Cardiovascular[6]	$27,400	
Hospital Admissions, Respiratory[6]	$21,000	
Ozone[7]		
Emergency Room Visits, Respiratory[5]	$369	
Hospital Admissions, Respiratory[6]	$16,400	$17,100

Notes: 1. Except for chronic bronchitis and acute myocardial infarction, medical expenditures per case are applied to the change in annual incidence. For chronic bronchitis and acute myocardial infarction, medical expenditures per case are applied to the annual changes in the prevalence of each disease, to generate an annual rather than lifetime estimate of costs for these chronic diseases.
2. Medical expenditure estimates for the following PM morbidity endpoints were not readily available: acute bronchitis, acute respiratory symptoms, asthma exacerbation, lower respiratory symptoms, upper respiratory symptoms, and work loss days.
3. Derived from Wittels et al. (1990) and Russell et al. (1998), both as cited in Abt Associates (2008).
4. Cropper and Krupnick (1999).
5. We assume that each E.R. visit equals one day of lost work time per worker affected. The estimate of 0.2 days per case reflects the percentage of cases realized by the working-age population, the ratio of workdays to total days in a year (235/365), and the percent of the working-age population in the labor force.
6. Agency for Healthcare Research and Quality (2000), as cited in Abt Associates (2008).
7. Medical expenditure estimates for the following ozone morbidity endpoints were not readily available: minor restricted activity days, school loss days, and outdoor worker productivity.

Medical Expenditures

To estimate the medical expenditures associated with changes in PM and ozone concentrations, the Project Team relied upon cost-of-illness estimates from the published literature. Table 8-4 presents the annual medical expenditures per case for those endpoints for which medical expenditure data were available. We applied the estimates presented in the table to the respective annual changes in incidence for each endpoint, except for chronic bronchitis and AMI. For these two endpoints, we applied the values from Table 8-4 to estimated changes in prevalence.

Summary of Benefit-Related Inputs

Table 8-5 summarizes the estimated changes in the labor force (i.e., the worker time endowment) associated with the Amendments for the 2010 and 2020 target years. Using the estimates in the table, the Project Team modified the time endowment for each model household included in EMPAX-CGE. The estimates in the table suggest that the U.S. labor force would be 0.34 percent smaller in 2010 and 0.57 percent smaller in 2020 if the Amendments had not been enacted. PM mortality effects would make up more than half of this reduction. Among morbidity endpoints, AMI and chronic bronchitis would have the most significant effect. The labor force impact of ozone pollution would represent less than five percent of the reduction in the labor force for each target year.

Table 8-5. Annual Change in Labor Force due to CAAA-Related Changes in Air Quality (Percent Change in Worker Time Endowment)

	2010	2020
Pollution-related Change in Worker Time Endowment	**0.34%**	**0.57%**
PM Mortality Subtotal	***0.18%***	***0.31%***
PM Morbidity Subtotal	***0.15%***	***0.25%***
Acute Myocardial Infarction	0.06%	0.09%
Chronic Bronchitis	0.05%	0.11%
Emergency Room Visits, Respiratory	<0.01%	<0.01%
Hospital Admissions, Cardiovascular	<0.01%	<0.01%
Hospital Admissions, Respiratory	<0.01%	<0.01%
Work Loss Days	0.04%	0.05%
Ozone Morbidity Subtotal	***0.01%***	***0.02%***
Emergency Room Visits, Respiratory	<0.01%	<0.01%
Hospital Admissions, Respiratory	<0.01%	<0.01%
Acute Respiratory Symptoms	<0.01%	<0.01%
School Loss Days	0.01%	0.01%
Worker Productivity	<0.01%	0.01%

Table 8-6. CAAA-Related Changes in Annual Medical Expenditures (Million 2006$).

	2010	2020
Pollution-related Change in Medical Expenditures	**$11,900**	**$19,600**
PM Morbidity Subtotal	***$11,600***	***$19,000***
Acute Myocardial Infarction	$9,500	$15,500
Chronic Bronchitis	$375	$919
Emergency Room Visits, Respiratory	$29	$39
Hospital Admissions, Cardiovascular	$1,228	$1,900
Hospital Admissions, Respiratory	$467	$683
Ozone Morbidity Subtotal	***$310***	***$580***
Emergency Room Visits, Respiratory	$2	$4
Hospital Admissions, Respiratory	$311	$575

Table 8-6 presents the estimated change in pollution-related medical expenditures associated with the Amendments. As indicated in the table, the Project Team estimates that medical expenditures related to air pollution would be approximately $12.9 billion higher in 2010 and $21 billion higher in 2020 in the absence of the Amendments. Similar to the labor force effects summarized in Table 8-5, PM-related morbidity, AMI in particular, represents most of the estimated change in pollution-related medical expenditures.

Empax-CGE Model Results

Using the inputs summarized in the previous section, the Project Team estimated the macroeconomic impacts of the Amendments under both the cost-only case and the labor force-adjusted case. As described above, the former captures the general equilibrium effects of CAAA compliance expenditures, whereas the latter accounts for the impacts of these expenditures as well as the labor force and medical expenditure impacts associated with the Amendments. We present the results of both analyses below.

Macroeconomic Impacts of CAAA Compliance Expenditures

Table 8-7 summarizes the results of the EMPAX-CGE cost-only model run. As the results in the table indicate, the Project Team estimates that the compliance expenditures associated with the Amendments will reduce GDP and consumption by approximately 0.5 percent in 2010 and 2020, relative to the without-CAAA scenario. The total estimated GDP reduction of $79 billion in 2010 and $110 billion in 2020 are 50 to 70 percent larger than the total primary cost estimates of $53 billion in 2010 and $65 billion in 2020. The difference is attributable to secondary effects of compliance costs on the overall economy, a large portion of which are likely the result of increases in energy prices, which has broad effects on overall production. Another factor is that investment in pollution control capital can divert capital from the purpose of enhancing long-term productivity within the industrial sector.

Table 8-7. Summary of Annual Macroeconomic Impacts: Cost-Only Case[1]

VARIABLE	MODEL RUN	2010	2015	2020
GDP	With Clean Air Act ($ billion)	$15,027	$17,338	$20,202
	Without Clean Air Act ($ billion)	$15,106	$17,430	$20,312
	Change ($ billion)	$79	-$93	$110
	% change	0.52%	-0.53%	0.54%
Consumption	With Clean Air Act ($ billion)	$10,969	$12,699	$14,881
	Without Clean Air Act ($ billion)	$11,023	$12,761	$14,956
	Change ($ billion)	$54	-$62	$75
	% change	0.49%	-0.49%	0.50%
Hicksian EV (annual)	Change ($ billion)	$54	-$62	$75
	% change	0.38%	-0.38%	0.39%

Notes: 1. Results are expressed in year 2006 dollars.

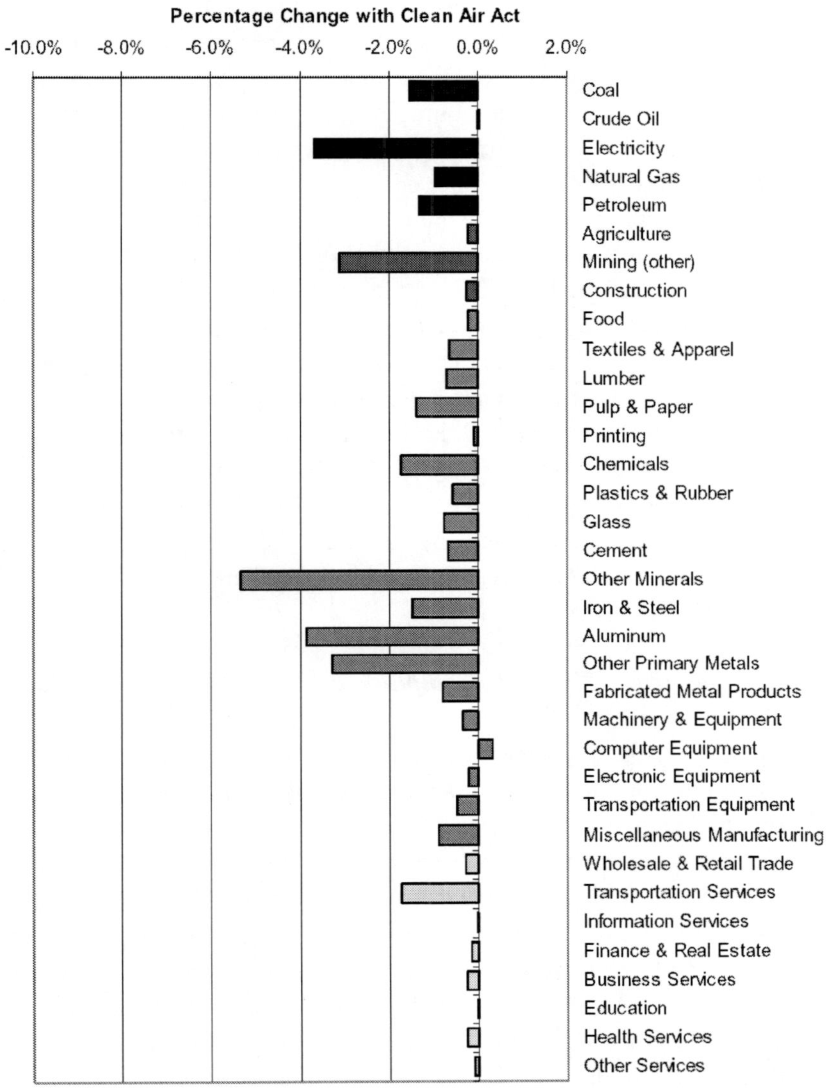

Figure 8-5. Percent Change in Industry Output in 2020: Cost-Only Case.

The percent reduction in equivalent variation is smaller than the corresponding reductions in GDP and consumption, at approximately 0.4 percent for both target years. This disconnect between the percent reduction in EV and the reductions in GDP and consumption suggests that, under the *with-CAAA* scenario, households allocate a greater share of their time endowment to leisure (and less to labor) than under the *without-CAAA* scenario. This increase in leisure partially offsets the welfare loss associated with reduced consumption.

Figure 8-5 presents the percent change in output by industry as estimated by EMPAX-CGE for the year 2020. The values in the table are typically highest among those industries with the most significant CAAA compliance expenditures relative to baseline industry revenue. For example, the electricity industry accounts for approximately 20 percent of CAAA compliance expenditures (approximately $14 billion, or 3.3 percent of benchmark electricity revenue); as a result, EMPAX-CGE estimates that output from the electricity industry declines by just less than 4 percent under the *with-CAAA* scenario relative to a U.S. economy without Clean Air Act programs. Because the power industry is the largest consumer of coal in the U.S., the reduction in output from the electricity industry also results in the secondary effect of reducing coal output by approximately 1.5 percent. The electricity industry's CAAA compliance expenditures also leads to higher electricity prices that prompt energy-intensive industries to switch to other energy sources (e.g., natural gas and oil) and/or seek energy efficiency improvements in their production process. In addition, because of CAAA requirements for cleaner (more expensive) fuels, petroleum sector output is projected to decline approximately 1.5 percent. The results in Figure 8-5 also suggest that the other minerals sector experiences the largest reduction in output, in proportional terms, among all industries (over 5 percent). This reflects the industry's high compliance expenditures relative to its size and the industry's energy-intensive production processes.

The industry-level results presented in Figure 8-5 also reflect the extent to which economic activity associated with CAAA compliance, such as new purchases of environmental protection goods and services, may partially offset the output losses associated with CAAA compliance expenditures. As a result of the CAAA, the demand for environmental protection goods and services will be higher relative to a U.S. economy without the Amendments.

Macroeconomic Impacts of CAAA Compliance Expenditures and Human Health Benefits

Building upon the results presented above, Table 8-8 summarizes the results of the EMPAX-CGE analysis for the labor force-adjusted case, which captures the full CAAA compliance expenditures as well as the labor force and medical expenditure benefits of the Amendments. The results presented in the table suggest that over time, the positive macroeconomic impacts of CAAA-related labor force and medical expenditure impacts slightly outweigh the negative macroeconomic effects of CAAA compliance costs. [115] For 2010, the results for the labor force-adjusted case show a reduction in GDP and consumption relative to the *without-CAAA* scenario, but the corresponding changes become positive in 2020. This largely reflects the rapid growth in the CAAA labor force effect between 2010 and 2020 (67 percent) relative to the growth in CAAA compliance expenditures (25 percent). We expect the CAAA-related labor force effect to grow more quickly than CAAA compliance expenditures during this period because, unlike compliance expenditures, the labor force effect is cumulative for the health endpoints with the most significant effect on the size of the labor force (i.e., premature mortality, chronic bronchitis, and AMI). In addition, the mortality effect is delayed relative to the time costs are incurred to reduce exposures because of the impact of the cessation lag.[116]

The results in Table 8-8 also suggest that the Amendments lead to an increase in household welfare, measured as the change in EV, under the labor force-adjusted case for both the 2010 and 2020 target years. The projected 0.8 percent *increase* in welfare for 2010 stands in contrast to the projected 0.21 percent *reduction* in GDP for that year and the 0.03 percent reduction in consumption. The fact that welfare rises while economic output declines indicates that, under the *with-CAAA* scenario, households allocate a greater share of their time endowment to leisure (and less to labor) than under the *without-CAAA* scenario. This reallocation of household time also occurs under the cost-only case, but it only partially offsets the negative welfare impact of reduced consumption. Under the labor force-adjusted case, the increase in leisure more than offsets the welfare loss associated with reduced consumption.

Figure 8-6 presents, by industry, the estimated percent change in output in 2020 for the labor force-adjusted case. The results in the figure indicate that, when labor force and medical expenditure impacts are accounted for, the CAAA leads to increased output for many industries and a decline in output for others. Consistent with the cost-only results, output in the computer equipment industry increases. The other sectors projected to experience an increase in output include many industries that tend to be labor-intensive and would benefit from a larger labor pool, such as most service industries. Output for health services declines, however, due to the reduction in health services demand associated with CAAA-related health improvements. Most of the other industries experiencing a reduction in output are either energy producers (e.g., electricity) or industries with energy-intensive production processes (e.g., iron and steel).

The results presented in Table 8-8 and Figure 8-6 show that the conclusions drawn from macroeconomic analyses of air pollution policy depend significantly on whether such analyses capture both the costs and at least some of the benefits of air policy. While the results of the EMPAX-CGE cost-only case suggest that the CAAA reduces the output of the U.S. economy by approximately 0.5 percent each year, the labor force-adjusted case shows that analyzing benefits in conjunction with costs can either reduce the magnitude or change the sign of model results. Therefore, general equilibrium analyses that examine cost-side macroeconomic impacts but ignore or overlook the impacts of policy-related labor force and health improvements may yield incomplete results that misinform policymakers and the public. The analysis presented in this chapter illustrates the feasibility of avoiding this outcome by examining both the costs and (a portion of the) benefits of air policy in a general equilibrium framework. It is important to note, however, that assessing expenditure-based output impacts should not replace the current practice of estimating the welfare (i.e., willingness-to-pay) benefits of avoided health effects. Unlike willingness-to-pay estimates, the results of CGE models do not reflect the non-market value that people place on avoided adverse health impacts. The outputs of such models represent a supplement to willingness-to-pay estimates rather than a substitute for such estimates.

Table 8-8. Summary of Annual Macroeconomic Impacts: Labor Force-Adjusted Case.[1]

VARIABLE	MODEL RUN	2010	2015	2020
GDP	With Clean Air Act ($ billion)	$15,027	$17,338	$20,202
	Without Clean Air Act ($ billion)	$15,059	$17,350	$20,197
	Change ($ billion)	$32	-$12	$5
	% change	0.21%	-0.07%	0.02%
Consumption	With Clean Air Act ($ billion)	$10,969	$12,699	$14,881
	Without Clean Air Act ($ billion)	$10,972	$12,696	$14,876
	Change ($ billion)	$3	$3	$5
	% change	0.03%	0.02%	0.03%
Hicksian EV (annual)	Change ($ billion)	$11	$22	$29
	% change	0.08%	0.13%	0.15%

Notes: 1. Results are expressed in year 2006 dollars.

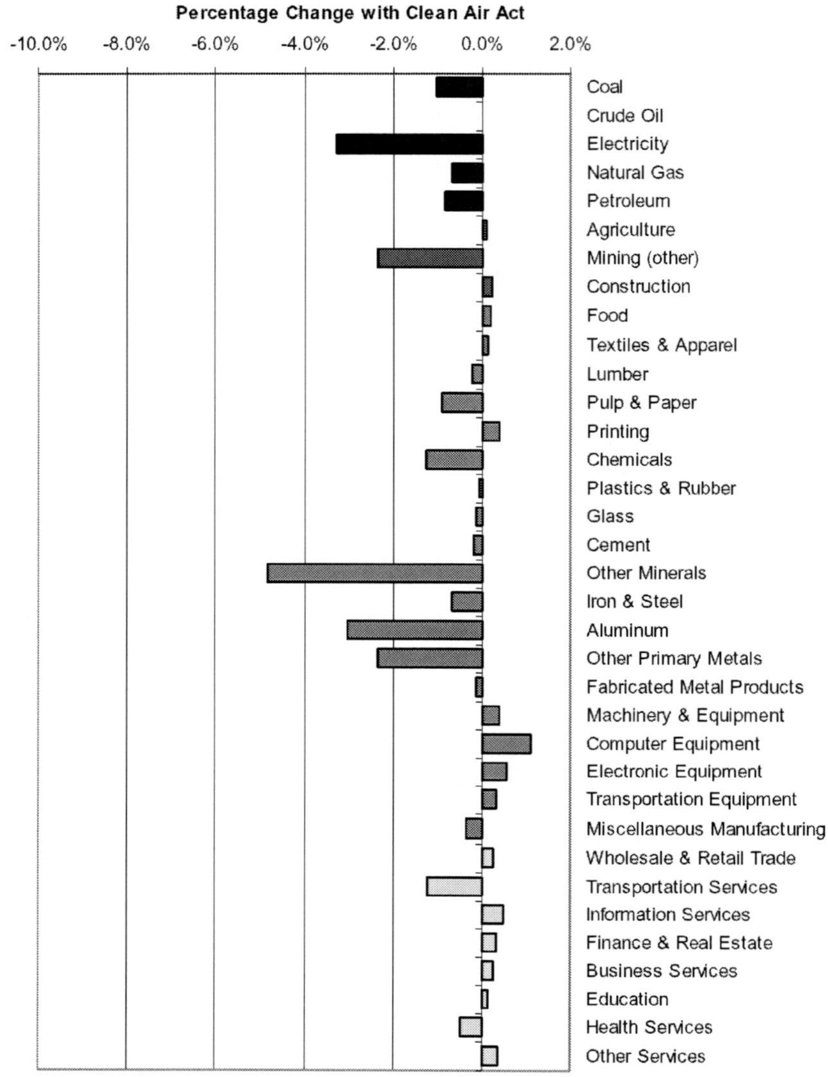

Figure 8-6. Percent Change in Industry Output in 2020: Labor Force-Adjusted Case.

Further work is needed, however, to reflect a much broader set of benefits in CGE models. As noted earlier, the results in this chapter are designed to supplement, but not replace, the more complete primary estimates of benefits and costs. The CGE model represents flows of products, labor, and capital between and among producers and consumers, but it excludes improvements in well-being due to enhanced longevity and health, except to the extent that these increase time available for labor and leisure among the workforce and reduce some medical costs. As a result, the vast majority of monetized benefits, many but not all of which represent benefits that are not traded in markets, cannot currently be reflected in a CGE model. This is the main reason that the beneficial results to the economy estimated in this chapter are substantially smaller than the primary estimate of benefits based on willingness to pay estimates. It is nonetheless important to realize that even the partial set of benefits-related impacts that are reflected in this chapter (i.e., labor force and medical expenditure impacts) more than compensate for the market costs we estimate to achieve CAAA compliance.

Analytic Limitations

While the analysis presented in this chapter provides a reasonable approximation of the macroeconomic impacts of the CAAA, we note the following limitations:

- ***Exclusion of labor force and leisure effects for individuals outside the formal economy:*** Given the uncertainty surrounding the macroeconomic impacts of retirees, children, and other populations who do not participate in formal labor markets and the fact that CGE models are ill-suited to address these uncertainties, the inputs developed by the Project Team for this analysis did not reflect changes in the time endowment for these individuals. To the extent that people outside formal labor markets contribute to the economy, we may underestimate the positive macroeconomic impacts of the Amendments.
- ***Exclusion of ozone mortality***: As described in the methods section, our analysis captures PM-related changes in mortality but does not account for mortality impacts from ozone exposure. Therefore, we likely underestimate the positive macroeconomic impacts of the Amendments.
- ***Exclusion of nonmarket and some market benefits:*** Our assessment of the macroeconomic impacts of the Amendments also excludes several other CAAA-related benefits that may improve economic performance or consumer welfare, such as visibility improvements, productivity enhancements in the agriculture and forestry sectors, reduced materials damage, and reduced pain and suffering from pollution-related illness. Because we do not capture these effects, we very likely underestimate the positive macroeconomic impacts of the Amendments.
- ***Assumption of separable benefits categories:*** Our modeling assumes labor supply and environmental quality are separable components of the utility function for households. This separability does not always hold, however; for example, cleaner air may encourage leisure activities such as birding and fishing, making air quality a complement to leisure. Prior work suggests that assuming separability may affect benefits by up to 30 percent in some cases.[117]
- ***Perfect foresight:*** EMPAX-CGE assumes that households have perfect foresight of future changes in policy and modify their current economic behavior accordingly. In reality, households often have imperfect information of future policy changes. Whether the assumption of perfect foresight leads to overestimation or underestimation of impacts is uncertain.
- ***EMPAX-CGE parameter uncertainty***: Similar to other CGE models, EMPAX-CGE requires the specification of several model parameters (e.g., elasticity values). Although the model relies upon credible values from the literature, the range of published estimates for many parameters varies widely across studies. It is uncertain whether the parameters included in EMPAX lead to overestimation or underestimation of impacts.

REFERENCES

[1] Abt Associates. (2009). Modeled Attainment Test Software User's Manual. Report prepared for Brian Timmon, US EPA Office of Air Quality Planning and Standards, Research Triangle Park, NC. March.

[2] Abt Associates (2008). Environmental Benefits Mapping and Analysis Program (BenMAP) User's Manual. Report Prepared for Office of Air Quality Planning and Standards, U.S. Environmental Protection Agency, Research Triangle Park, NC, Neal Fann, Project Manager.

[3] Alberini, A., Cropper, M., Krupnick, A. & Simon, N. B. (2004). "Does the Value of a Statistical Life Vary with Age and Health Status? Evidence from the U.S. and Canada". *Journal of Environmental Economics and Management, 48(1),* 769-792.

[4] Aldy, Joseph E. & Kip Viscusi, W. (2007). "Age Differences in the Value of Statistical Life: Revealed Preference Evidence." *Rev Environ Econ Policy, 1*, 241-260.

[5] Barbera, A. J. & McConnell, V. D. (1986) "Effects of Pollution Control on Industry Productivity: A Factor Demand Approach." *The Journal of Industrial Economics*, Vol. *XXXV*, 161-172.

[6] Barbera, A. J. & McConnell, V. D. (1990) "The Impact of Environmental Regulations on Industry Productivity: Direct and Indirect Effects." *Journal of Environmental Economics and Management*, Vol. *18*, 50-65.

[7] Bell, M. L., Dominici, F. & Samet, J. M. (2005). A meta-analysis of time-series studies of ozone and mortality with comparison to the national morbidity, mortality, and air pollution study. *Epidemiology*, Vol, *16 (4)*, 436-45.

[8] Bell, M. L., et al. (2004). Ozone and short-term mortality in 95 US urban communities, 1987-2000. *JAMA*, *292(19),* 2372-8.

[9] Byun, D. W. & Ching, J. K. S. (1999). "Science Algorithms of the EPA Models-3 Community Multiscale Air Quality (CMAQ) Modeling System." U.S. EPA Office of Research and Development, Washington, D.C. (EPA/600/R-99/030).

[10] Carbone, J. C. & Smith, V. K. (2008). "Evaluating policy interventions with general equilibrium externalities." *Journal of Public Economics*, *92*, 1254-1274.

[11] Carlson, Curtis P., Dallas Burtraw, Maureen Cropper & Karen Palmer. (2000) "SO_2 Control by Electric Utilities: What are the Gains from Trade?" *Journal of Political Economy*, *108 (6)*, 1292-1326.

[12] Chestnut, L. G., R. D. Rowe & Breffle, W.S. (2004). *Economic Valuation of Mortality Risk Reduction: Stated Preference Approach in Canada.* Report prepared for Paul De Civita, Health Canada by Stratus Consulting Inc., Boulder, CO, December.

[13] Committee on Vehicle Emission Inspection and Maintenance Programs, Board on Environmental Studies and Toxicology, Transportation Research Board, National Research Council. (2001). Evaluating Vehicle Emissions Inspection and Maintenance Programs.

[14] Crawford, S. (2009). Personal communication on September 4, 2009 and September 15, 2009. New York State Department of Environmental Conservation, Division of Lands and Forests, Forest Utilization Program.

[15] Crump, K. S. (1994). Risk of benzene-induced leukemia: A sensitivity analysis of the Pliofilm Cohort with additional follow-up and new exposure estimates. *Journal of Toxicology and Environmental Health, 42*, 219-242.

[16] DeShazo, J.R., and T.A. Cameron. (2004). "Mortality and Morbidity Risk Reduction: An Empirical Life-Cycle Model of Demand with Two Types of Age Effects." Unpublished paper, Department of Policy Studies, University of California at Los Angeles.

[17] Driscoll, Charles T. et al. (2003). Chemical Response of Lakes in the Adirondack Region of New York to Declines in Acidic Deposition. *Environmental Science and Technology, 37(10)*, 2036-2042.

[18] Driscoll, Charles T. et. al. (2001). Acidic Deposition in the Northeastern United States: Sources and Inputs, Ecosystem Effects, and Management Strategies. *BioScience, 51(3)*, 180-198.

[19] Dutton, John M. & Annie Thomas, (1984). "Treating Progress Functions as a Managerial Opportunity," *Academy of Management Review*, *9(2)*, 235-247.

[20] Eftim, S. E., Samet, J. M., Janes, H., McDermott, A. & Dominici, F. (2008). Fine particulate matter and mortality: a comparison of the six cities and American Cancer Society cohorts with a medicare cohort. *Epidemiology, 19(2)*, 209-216.

[21] Ellerman, A. Denny. (2003). "Ex Post Evaluation of Tradable Permits: The U.S. SO_2 Cap-and-Trade Program," MIT CEEPR Working Paper WP-2003-003, available at: web.mit.edu/ceepr/www/publications/workingpapers_2000_2004.html#2003.

[22] Ellerman, A. Denny, Paul L. Joskow, Richard Schmalensee, Juan-Pablo Montero & Elizabeth Bailey (2000). *Markets for Clean Air: The U.S. Acid Rain Program*. Cambridge University Press.

[23] Epple, Dennis, Linda Argote & Rukmini Devadas, (1991). "Organizational Learning Curves: A Method for Investigating Intra-plant Transfer of Knowledge Acquired Through Learning by Doing," *Organizational Science*, *2(1)*, February 1991.

[24] Gray, W. B. & Shadbegian, R. J. (1994). "Pollution Abatement Costs, Regulation, and Plant-Level Productivity." Center for Economic Studies.

[25] Hammitt, James K. (2007). "Valuing Changes in Mortality Risk: Lives Saved Versus Life Years Saved." *Rev Environ Econ Policy, 1*, 228-240.

[26] Harrington, Winston, Robert D. Morgenstern & Peter Nelson. (1999). "On the Accuracy of Regulatory Cost Estimates," Resources for the Future Discussion Paper 99-18, January 1999.

[27] Huang, Y., Dominici, F. & Bell, M. L. (2005). Bayesian hierarchical distributed lag models for summer ozone exposure and cardio-respiratory mortality. *Environmetrics.* Vol, *16*, 547–562.

[28] Hubbell, Brian J., Richard V. Crume, Dale M. Evarts & Jeff M. Cohen. (2010). "Regulation and Progress under the 1990 Clean Air Act Amendments" *Review of Environmental Economics and Policy* 4(1): 122-138.

[29] ICF International. (2007). The HAPEM6 User's Guide: Hazardous Air Pollutant Exposure Model, Version 6. Prepared for the Office of Air Quality Planning and Standards, US EPA, Research Triangle Park, NC.

[30] Industrial Economics, Inc. (2006). Expanded Expert Judgment Study of the Concentration-Response Relationship Between $PM_{2.5}$ Exposure and Mortality. Prepared

for the Office of Air Quality Planning and Standards, U.S. Environmental Protection Agency, September.

[31] Industrial Economics, Inc. (2009). Section 812 Prospective Study of the Benefits and Costs of the Clean Air Act: Air Toxics Case Study – Health Benefits of Benzene Reductions in Houston, 1990-2020. Prepared for the Office of Policy Analysis and Review, U.S. Environmental Protection Agency, Washington, DC. July.

[32] International Energy Agency, (2000). "Experience Curves for Energy Technology Policy"

[33] Ito, K., S. F. De Leon and M. Lippmann, 2005. Associations between ozone and daily mortality: analysis and meta-analysis. *Epidemiology*. Vol. 16 (4): 446-57.

[34] Jerrett, M., R.T. Burnett, et al. (2005). Spatial analysis of air pollution and mortality in Los Angeles. *Epidemiology* 16(6): 1-10. Jerrett, M.J., R.T. Burnett, C. Arden Pope III, K. Ito, G. Thurston, D. Krewski, Y. Shi, E. Calle, M. Thun. (2009). Long-Term Ozone Exposure and Mortality. *New England Journal of Medicine* 360: 1085-95.

[35] Johansson, P.-O. (2002). "On the definition and age dependency of the value of a statistical life." *Journal of Risk and Uncertainty*. 25: 251-63.

[36] Jones-Lee, M.W. (1992). "Paternalistic Altruism and the Value of Statistical Life." *The Economic Journal.* 102:80-90. January.

[37] Jorgenson, Dale W., Richard J. Goettle, Brian H. Hurd, and Joel B. Smith, et al. (2004). U.S. Market Consequences of Global Climate Change, prepared for the Pew Center on Global Climate Change, April 2004.

[38] Joskow, Paul L. and Nancy L. Rose, (1985) "The Effects of Technological Change, Experience, and Environmental Regulation on the Construction Cost of Coal-Burning Generating Units," *RAND Journal of Economics*, 16(1): 1-27.

[39] Kniesner, Thomas J., W. Kip Viscusi, and James P. Ziliak, (2010) "Policy relevant heterogeneity in the value of statistical life: New evidence from panel data quantile regressions," *Journal of Risk and Uncertainty* 40:15–31.

[40] Krewski, D., M. Jerrett, R.T. Burnett, et al. (2009). Extended follow-up and spatial analysis of the American Cancer Society study linking particulate air pollution and mortality. *Res Rep Health Eff Inst.* 140: 5-114.

[41] Krewski, D., R.T. Burnett, et al. (2000). Reanalysis of the Harvard Six Cities Study and the American Cancer Society Study of Particulate Air Pollution and Mortality. Special Report to the Health Effects Institute, Boston MA, July.

[42] Krupnick, Alan. (2007). "Mortality-risk Valuation and Age: Stated Preference Evidence." *Rev Environ Econ Policy* 1: 261-282. Krupnick, A.J. and M.L. Cropper (1992). The effect of information on health risk valuations. *Journal of Risk and Uncertainty* 5(1): 29-48.

[43] Kunzli, N., S. Medina, et al. (2001). Assessment of deaths attributable to air pollution: should we use risk estimates based on time series or cohort studies? *American Journal of Epidemiology* 153(11): 1050-1055.

[44] Laden, F., J. Schwartz, et al. (2006). Reduction in Fine Particulate Air Pollution and Mortality: Extended Follow-up of the Harvard Six Cities Study. *American Journal of Respiratory and Critical Care Medicine* 173: 667-672.

[45] Lee, E.H. and W.E. Hogsett. (1996). Methodology for Calculating Inputs for Ozone Secondary Standard Benefits Analysis: Part II. Prepared for the U.S. EPA, Office of Air Quality Planning and Standards, Air Quality Strategies and Standards Division.

[46] Levy, J. I., S. M. Chemerynski and J. A. Sarnat, 2005. Ozone exposure and mortality: an empiric bayes metaregression analysis. *Epidemiology*. Vol. 16 (4): 458-68.
[47] McLaughlin, D. (1998). A decade of forest tree monitoring in Canada: evidence of air pollution effects. *Environ. Rev*. 6(3-4):151-171 Magat, W.A., Viscusi, W.K., et al. (1996). A reference lottery metric for valuing health. Management Science 42:1118-1130.
[48] Morgenstern, R.D., Pizer, W.A., and Shih, J-S. (1998) "The Cost of Environmental Protection." Discussion Paper 98-36. Resources for the Future.
[49] National Research Council (NRC) (2008). Estimating Mortality Risk Reduction and Economic Benefits from Controlling Ozone Air Pollution. The National Academies Press, Washington, DC.
[50] New York State Department of Environmental Conservation. (2009). Stumpage Price Report (Winter 2009/#74). New York State Department of Environmental Conservation, Division of Lands and Forests, Forest Utilization Program. Albany, New York.
[51] Pope, C. A., R. T. Burnett, et al. (2002). Lung cancer, cardiopulmonary mortality, and long-term exposure to fine particulate air pollution. *Journal of the American Medical Association* 287(9): 1132-1141.
[52] Puett, R.C., J. Schwartz, J.E. Hart, J.D. Yanosky, F.E. Speizer, H. Suh, C.J. Paciorek, L.M. Neas, and F. Laden. 2008. Chronic particulate exposure, mortality, and coronary heart disease in the nurses' health study. *Am J Epidemiol.* 168(10):1161-1168.
[53] Puett, R.C., J. Hart, J.D. Yanosky, C. Paciorek, J. Schwartz, H. Suh, F.E. Speizer, and F. Laden. 2009. Chronic Fine and Coarse Particulate Exposure, Mortality, and Coronary Heart Disease in the Nurses' Health Study. *Environmental Health Perspectives* 117:1697-1701.
[54] Rea, A., J. Lynch, R. White, G. Tennant, J. Phelan and N. Possiel. 2009. Critical Loads as a Policy Tool: Highlights of the NOx/SOx Secondary National Ambient Air Quality Standard Review. Slide 6: Nationwide Total Reactive Nitrogen Deposition (2002). Available online at: http://nadp.sws.uiuc.edu/meetings/fall2009/post/session4.html.
[55] Rowlatt, Penelope, Michael Spackman, Sion Jones, Michael Jones-Lee, and Graham Loomes. (1998). Valuation of Deaths from Air Pollution. For the Department of Environment, Transport and the Regions and the Department of Trade and Industry. February.
[56] RTI International. (2008). EMPAX-CGE Model Documentation, prepared for U.S. EPA Office of Air Quality Planning and Standards, March 2008.
[57] Samet, J. M., F. Dominici, et al. (2000). The National Morbidity, Mortality, and Air Pollution Study Part I: Methods and Methodologic Issues. Research Report 94, Health Effects Institute, Boston, MA.
[58] Schwartz, J., (2005). How sensitive is the association between ozone and daily deaths to control for temperature? *Am J Respir Crit Care Med*. Vol. 171 (6): 627-31.
[59] Schwartz, J., B. Coull, F. Laden, L. Ryan. (2008). The effect of dose and timing of dose on the association between airborne particles and survival. Environmental Health Perspectives 116(1): 64-9.
[60] U.S. Department of Agriculture, Forest Service, Rocky Mountain Region. (2000) Screening Methodology for Calculating ANC Change to High Elevation Lakes: User's Guide. January 2000.

[61] U.S. Department of Energy, Energy Information Administration, (2005). Annual Energy Outlook 2005.

[62] U.S. Environmental Protection Agency (2006). Regulatory Impact Analysis for the 2006 National Ambient Air Quality Standards for Particle Pollution. http://www.epa.gov/ttn/ecas/ria.html

[63] U.S. Environmental Protection Agency (2004). User's Guide for the AMS/EPA Regulatory Model – AERMOD. Office of Air Quality Planning and Standards, Research Triangle Park, NC, Report No. EPA-454/B-03-001. http://www.epa.gov/scram001/7thconf/aermod/aermodugb.pdf

[64] U.S. Environmental Protection Agency (2003). Response of surface water chemistry to the Clean Air Act Amendments of 1990. EPA 620/R-03/001.

[65] U.S. Environmental Protection Agency (2000). Guidelines for Preparing Economic Analyses, EPA 240-R-00-003, September.

[66] U.S. Environmental Protection Agency (1999). Cost of Illness Handbook. http://www.epa.gov/oppt/coi/pubs/I_1.pdf.

[67] U.S. Environmental Protection Agency (1998). Carcinogenic Effects of Benzene: An Update. Office of Research and Development, Washington, DC. EPA/600/P-97/001F.

[68] U.S. Environmental Protection Agency, Science Advisory Board (2007). "Benefits and Costs of Clean Air Act – Direct Costs and Uncertainty Analysis", Advisory Letter, June 8, 2007. Washington, DC. EPA-SAB-COUNCIL-ADV-07-002, Available at: http://www.epa.gov/sab/pdf/council-07-002.pdf.

[69] U.S. Environmental Protection Agency, Science Advisory Board (2004). "Advisory on Plans for Health Effects Analysis in the Analytical Plan for EPA's Second Prospective Analysis—Benefits and Costs of the Clean Air Act, 1990-2020: Advisory by the Health Effects Subcommittee of the Advisory Council on Clean Air Compliance Analysis." Washington, DC. EPA-SAB-COUNCIL-ADV-04-002.

[70] U.S. Environmental Protection Agency, Science Advisory Board (2001). "Review Of The Draft Analytical Plan For EPA's Second Prospective Analysis - Benefits And Costs Of The Clean Air Act 1990-2020". Washington, D.C. EPA-SAB-COUNCIL-ADV-01-004.

[71] The U.S. National Acid Precipitation Assessment Program. 1991. Integrated Assessment Report. The NAPAP Office of the Director, Washington, DC.

[72] U.S. Office of Management and Budget (2003). Regulatory Analysis. OMB Circular A-4, September 17. http://www.whitehouse.gov/OMB/circulars/a004/a-4.pdf.

[73] Viscusi, W.K. (1992). *Fatal Tradeoffs*, (Oxford University Press: New York), Table 4-1.

[74] Viscusi, W. K. and J. E. Aldy. 2003. The Value of a Statistical Life: A Critical Review of Market Estimates throughout the World. AEI-Brookings Joint Center for Regulatory Studies. Washington, DC. January.

[75] Viscusi, W.K. (2004). The value of life: Estimates with risks by occupation and industry. *Economic Inquiry* 42(1): 29-48.

[76] Wilson, J. H., M. A. Mullen, A. D. Bollman, K. B. Thesing, M. Salhotra, F. Divita, J. Neumann, J. C. Price, and J. DeMocker. (2008). Emissions projections for the U.S. Environmental Protection Agency Section 812 second prospective Clean Air Act cost/benefit analysis. *J. of Air & Waste Manage. Assoc.* 58:657-672.

[77] Woods & Poole Economics Inc., 2007. Complete Demographic Database. Washington, DC. http://woodsandpoole.com/index.php.

[78] Zeger, S.L., F. Dominici, A. McDermott, and J.M. Samet. 2008. Mortality in the Medicare population and chronic exposure to fine particulate air pollution in urban centers (2000-2005). *Environ Health Perspect* 116(12):1614-1619.

End Notes

[1] The lone exception is the Coke Ovens Residual Risk rulemaking, promulgated under Title III of the Act in March 2005. We omitted this rule because it has a very small impact on criteria pollutant emissions (less than 10 tons per year VOCs) relative to the overall impact of the CAAA. The primary MACT rule for coke oven emissions, however, involves much larger reductions and therefore is included in the with-CAAA scenario.

[2] See_www.epa.gov/oar/sect812 for a complete list and opportunity to download supporting documentation for this Second Prospective analysis.

[3] Note that some elements of Title VII enforcement efforts, such as settlements for historical violations of CAA provisions, particularly in the electric utility and petroleum refining sectors, are included in the emissions inventories of the with-CAAA scenario. For more information, see EPA's detailed emissions report supporting this study at www.epa.gov/oar/sect812

[4] Implementing this approach has occasionally required some difficult decisions on what constitutes 1990 levels of emissions controls. In general, we have interpreted any rules that were promulgated as final prior to 1990 to be part of *the withoutCAAA* scenario baseline. The residential wood stove New Source Performance Standard, however, was promulgated in 1988, but is not part of the *without-CAAA* scenario, because EPA did not certify NSPS compliant wood stoves until 1992. In this case, perhaps incorrectly, we interpreted the effective date of 1992 as the determining factor in whether the level of emissions stringency in 1990 should include the wood stove NSPS.

[5] Area sources are also commonly referred to as nonpoint sources. We estimated utility and industrial point source emissions at the plant/facility level. We estimated nonroad engine/vehicle, motor vehicle, and area source emissions at the county level.

[6] The exceptions are where 1990 emissions were re-computed using updated methods developed for the 2002 National Emissions Inventory (NEI) for selected source categories with the largest criteria pollutant emissions and most significant methods changes.

[7] We used the draft NEI point source file because the final version of that file was not available at the time the analysis was performed. For area sources, we used the final NEI nonpoint source file.

[8] Due to resource constraints and model limitations, we relied primarily on a validation analysis EPA conducted on 2001 emissions, rather than developing a new analysis for the year 2000.

[9] These emissions sources include agricultural production-crops, fertilizer application, and nitrogen solutions; agricultural tilling; animal husbandry; aircraft; forest wildfires; prescribed burning for forest management; residential wood fireplaces and wood stoves; and unpaved roads.

[10] The table presents 2002 data in order to be consistent with EPA's 2002 NEI, which we used to estimate emissions from industrial point sources and area sources.

[11] One exception is the Coke Ovens Residual Risk rulemaking, promulgated under Title III of the Act in March 2005. We omitted this rule because it has a very small impact on criteria pollutant emissions (less than 10 tons per year VOCs) relative to the *with-CAAA* scenario. The primary Maximum Achievable Control Technology (MACT) rule for coke oven emissions, however, involves much larger reductions and therefore is included in the *with-CAAA* scenario. In addition, we also modeled emissions reductions from local controls implemented to comply with the 8-hour Ozone NAAQS, the $PM_{2.5}$ NAAQS, and the Clean Air Visibility Rule, using the proposed or promulgated forms of these rules as of January 2008.

[12] Under Phase II of the OTC memorandum of understanding, eleven eastern states committed themselves to achieving regional reductions in NO_x emissions through a cap-and-trade system similar to the SO_2 trading program established under Title IV of the amendments.

[13] As noted earlier in this chapter, the woodstove NSPS was interpreted as part of the differential between the with- and without-CAAA scenarios. NSPS compliance is required only for new units, which in practice are replaced very slowly. We estimate that, almost 20 years after NSPS implementation, in 2010, about 70 percent of the wood stoves in use are preNSPS uncertified models; by 2020, we estimate that turnover will reduce non-certified unit usage to just under 65 percent.

[14] U.S. Environmental Protection Agency. Clean Air Markets – Data and Maps_≤http://camddataandmaps.epa.gov/gdm/≥ Accessed March 2009.

[15] Dr. A. Denny Ellerman's approach relies on multiplying a "baseline" pre-Title IV emissions rate by 2001 CEM heat input observations for each electric generating unit.

[16] Private costs differ from the direct cost estimates presented in this chapter in two important ways: (1) they reflect private interest rates rather than the 5 percent social discount rate used throughout this report and (2) they reflect transfers (e.g., excise taxes on fuel) not included in our direct cost estimates.

[17] A full list of the regulations incorporated in the *with-CAAA* scenario is presented in Table 2-3.

[18] Except to the extent they are co-controlled by VOC limits or other measures focused on criteria pollutants, reductions in emissions of hazardous air pollutants were omitted because our benefits analysis focuses on the effect of criteria pollutants. Benefits of HAP emissions reductions are discussed in the context of a limited case study, however, in Chapter 5 of this document. In addition, no CAAA emissions control measures are currently targeted to control NH_3 emissions, so no costs for NH_3 control are included in our overall CAAA cost estimates.

[19] See the report, Direct Cost Estimates for the Clean Air Act Second Section 812 Prospective Analysis. Available at www.epa.gov/oar/sect812.

[20] For PM NAAQS compliance, an optimization approach was not possible, because target emissions reductions were not available for each non-attainment area. Instead, we developed a model SIP for all PM nonattainment areas, and estimated costs for those measures in the model SIP for each nonattainment area.

[21] These studies include John M. Dutton and Annie Thomas, "Treating Progress Functions as a Managerial Opportunity," Academy of Management Review, 1984, Vol. 9, No. 2, 235-247; Dennis Epple, Linda Argote, and Rukmini Devadas, "Organizational Learning Curves: A Method for Investigating Intra-plant Transfer of Knowledge Acquired Through Learning by Doing," Organizational Science, Vol. 2, No. 1, February 1991; International Energy Agency, Experience Curves for Energy Technology Policy, 2000; and Paul L. Joskow and Nancy L. Rose, "The Effects of Technological Change, Experience, and Environmental Regulation on the Construction Cost of Coal-Burning Generating Units," RAND Journal of Economics, Vol. 16, Issue 1, 1-27, 1985.

[22] The Council recommended that we apply a default learning rate of 5 to 10 percent to sectors for which no empirical data are available. We chose 10 percent as a default learning rate because this value is more consistent with the learning rates presented in the empirical literature than the low end of the Council's recommended range.

[23] The Project Team makes no learning curve adjustments for motor vehicle inspection and maintenance programs. Because most states either run centralized inspection centers themselves or regulate the fees charged by decentralized inspection centers, it is unclear whether the learning curve impacts for I&M programs would be significant.

[24] Note that the discount rate we use to annualize capital investment costs is distinct from the discount rate used to calculate the total net present value of costs and benefits incurred through the full 1990 to 2020 study period. The net present value of costs and benefits is examined separately in Chapter 7 where we compare total costs to total benefits.

[25] See, for example, A Denny Ellerman, 2003, "Ex Post Evaluation of Tradable Permits: The U.S. SO_2 Cap-and-Trade Program," MIT Center for Energy and Environmental Policy Research Working Paper number WP-2003-003, available at: web. Ellerman cites two papers for these estimates: Curtis P. Carlson, Dallas Burtraw, Maureen Cropper, and Karen Palmer, (2000) "SO_2 Control by Electric Utilities: What are the Gains from Trade?" Journal of Political Economy, 108 (6):1292-1326; and A. Denny Ellerman, Paul L. Joskow, Richard Schmalensee, Juan-Pablo Montero, and Elizabeth Bailey (2000). Markets for Clean Air: The U.S. Acid Rain Program. Cambridge University Press.

[26] The NAPAP assessment cites a range of $1 billion to $2 billion, in year 2000 dollars. Adjusting for inflation using the GDP deflator, this range increases to $1.2 billion to $2.3 billion in year 2006 dollars.

[27] We present sensitivity test results for 2020 estimates because the differences between the primary cost estimates and the alternative cost estimates discussed in this section are most pronounced in 2020.

[28] Note that in both our central case estimates and in our sensitivity analysis for fleet composition, the same fleet composition is assumed in the *with-CAAA* and *without -CAAA* scenarios. It is likely that, as compliance costs increase, the CAAA could have a significant effect on fleet composition, but our current analysis does not address that factor.

[29] Committee on Vehicle Emission Inspection and Maintenance Programs, Board on Environmental Studies and Toxicology, Transportation Research Board, National Research Council. *Evaluating Vehicle Emissions Inspection and Maintenance Programs*. 2001.

[30] For an analysis of the learning rates estimated in the empirical literature, see John M. Dutton and Annie Thomas, "Treating Progress Functions as a Managerial Opportunity," *Academy of Management Review*, Vol 9, No. 2, 1984.

[31] Barbera, A.J. and McConnell, V.D. (1986) "Effects of Pollution Control on Industry Productivity: A Factor Demand Approach." *The Journal of Industrial Economics*. Vol. XXXV, 161-172.

Barbera, A.J. and McConnell, V.D. (1990) "The Impact of Environmental Regulations on Industry Productivity: Direct and Indirect Effects." *Journal of Environmental Economics and Management*. Vol. 18, 50-65.

Gray, W.B. and Shadbegian, R.J. (1994) "Pollution Abatement Costs, Regulation, and Plant-Level Productivity." Center for Economic Studies.

Morgenstern, R.D., Pizer, W.A., and Shih, J-S. (2001) "The Cost of Environmental Protection." *Review of Economics and Statistics* Vol. 83, No. 4, 732-738. (doi:10.1162/003465301753237812).

[32] Barbera and McConnell (1986) found a negative impact of pollution control on productivity, while Barbera and McConnell (1990) and Gray and Shadbegian (1994) found an ambiguous impact, and Morgenstern et al. (1998) found a positive impact.

[33] The latest available version of CMAQ, version 4.6, was used for this study. This version of the model supports several different gas-phase chemical mechanism, particle treatment, aerosol deposition, and cloud treatment options. All simulations conducted as part of this study used the CB05 chemical mechanism. For particles, the AERO4 particle treatment, which includes sea salt, was applied. Finally, the plume-in-grid feature of CMAQ was not used for this study. More details are available in *Second Prospective Analysis of Air Quality in the U.S.: Air Quality Modeling*, available at www.epa.gov/oar/sect812

[34] Memorandum of June 14, 2010 to Jim DeMocker, EPA, from Tyra Walsh, Henry Roman, and Jim Neumann, Industrial Economics, Inc. (IEc), "Description of the Adjustment to the Primary Particulate Matter Emissions Estimates and the Modeled Attainment Test Software Analysis (MATS) Procedure for the 812 Second Prospective Analysis." The memo is available at www.epa.gov/oar/sect812.

[35] Most FRM monitors (about 75 percent) are not co-located with a speciation monitor. Therefore, we also used data providing speciated PM mass from the STN and IMPROVE monitors. The MATS analysis used speciated data from 273 STN or IMPROVE monitors with at least two valid quarters of speciated data in 2002.

[36] The STN bar charts include an estimated water component, which the MATS input monitor files include to make STN and IMPROVE monitor data consistent with FRM monitor data. The water component is not an STN component, but was estimated using the SANDWICH (Sulfates, Adjusted Nitrates, Derived Water, Inferred Carbonaceous mass, and estimated aerosol acidity (H+)) process.

[37] There is one area in northeastern Utah where the MATS procedure yields results for the *without-CAAA* scenario that are so large as to be not plausible. The result was associated with increases in agricultural burning in the *without-CAA* scenario, coupled with otherwise low organic carbon monitor values in nearby monitors – the application of MATS therefore led to unusually high organic carbon and $PM_{2.5}$ measures for that area. For those three cells, we performed a moving average smoothing procedure to re-estimate the without-CAAA concentrations, using PM estimates from adjoining cells. The adjustment is used only for the purposes of generating the maps in this chapter; for the purposes of health benefits modeling and valuation of benefits, we excluded these three suspect cells. The cells represent very rural, sparsely populated areas in the Wasatch Mountains, and so we believe that excluding them from the benefits calculations is both prudent and has only a modest underestimation effect on the overall health benefits estimates.

[38] We examined this result further and found that, in cells with the largest disbenefits, the 2020 *without -CAAA* scenario yields concentrations of approximately 45 ppb, while concentrations in outlying areas are as high as 100 ppb or slightly higher. One effect of CAAA controls is to suppress NO_x-scavenging in the city center, where disbenefits are largest, yielding *with-CAAA* concentrations in the 60 to 65 ppb range. The main effect of the CAAA, however, is large decreases in ozone in the outlying areas, to concentrations of 60 to 75 ppb. The net effect on a population weighted basis remains a lowering of overall exposures.

[39] Use of an integrated model such as CMAQ for the current study was one of the recommendations made by the Council in their review of the First Prospective analysis.

[40] ICF International, Evaluation of CMAQ Model Performance for the 812 Prospective II Study, November 24, 2009, page 31

[41] For more information, see the BenMAP *User's Manual and Appendices*, September 2008, Prepared for the Office of Air Quality Planning and Standards, U.S. Environmental Protection Agency, Research Triangle Park, NC, by Abt Associates Inc.

[42] As noted in Chapter 4, eVNA and VNA are procedures for interpolating values from nearby monitors using inverse distance squared weighting using Voronoi Neighbor Averaging. This is an algorithm that identifies a set of monitors close to the grid cell (called "neighbors") and then estimates the PM species concentration in that grid cell by calculating an inverse- distance weighted average of the monitor values (i.e., the concentration values at monitors closer to the grid cell are weighted more heavily than monitors that are further away). See the BenMAP manual for further information on the eVNA procedure. Abt Associations (2008). *BenMAP: Environmental Benefits Mapping and Analysis Program User's Manual*. Prepared for the U.S. Environmental Protection Agency's Office of Air Quality Planning and Standards, Research Triangle Park, NC, September.

[43] Woods & Poole Economics Inc., 2007. Complete Demographic Database. Washington, DC. http://woods andpoole.com/index.php.

[44] See *Health and Welfare Benefits Analyses to Support the Second Section 812 Benefit-Cost Analysis of the Clean Air Act*, February 2011, for a list of data sources and average baseline incidence rates for each health effect.

[45] See Kunzli et al. (2001) for a discussion of this issue.

[46] Studies have shown that greater spatial resolution of exposures can result in increased effect estimates (Jerrett et al., 2005).

[47] See "Health Effects Subcommittee of the Council. Review of EPA's Draft Health Benefits of the Second Section 812 Prospective Study of the Clean Air Act." (EPA-COUNCIL-10-001), available at http://www.epa.gov/advisorycouncilcaa

[48] Note that these results are based on a slightly different air quality dataset than the analysis of the full cohort. The nationwide estimate is based on a six-year average (2000-2005) and the ACS and Six Cities location-specific results are based on two years of data (2000-2002).

[49] Biennial questionnaires on lifestyle factors were administered to participants, allowing for control of a number of individual-level confounders.

[50] For specific examples of research addressing differential toxicity of PM components, see Chapter 5 of *Uncertainty Analyses to Support the Second Section 812 Benefit-Cost Analysis of the Clean Air Act.* http://www.epa.gov/oar/sect812/may10/IEc_Uncertainty.pdf

[51] Note that data presented in Table 5-3 are for the most important $PM_{2.5}$ components; some less important species, with lower concentrations, are omitted.

[52] See, for example, National Research Council, 2008, *Estimating Mortality Risk Reduction and Economic Benefits from Controlling Ozone Air Pollution*. A key recommendation of this NAS panel was that ozone mortality estimates from available epidemiological studies represent a separate and additive effect to those from PM/mortality epidemiological studies.

[53] National Research Council (NRC) (2008). Estimating Mortality Risk Reduction and Economic Benefits from Controlling Ozone Air Pollution. The National Academies Press, Washington, DC.

[54] See "Health Effects Subcommittee of the Council. Review of EPA's Draft Health Benefits of the Second Section 812 Prospective Study of the Clean Air Act." (EPA-COUNCIL-10-001), available at http://www.epa.gov/advisorycouncilcaa

[55] Because people are valuing small decreases in the risk of premature mortality, it is expected deaths that are inferred. For example, suppose that a given reduction in pollution confers on each exposed individual a decrease in mortal risk of 1/100,000. Then among 100,000 such individuals, one fewer individual can be expected to die prematurely. If the average individual's WTP for that risk reduction is $50, then the implied value of a statistical premature death avoided in that population is $50 x 100,000 = $5 million.

[56] For a more detailed discussion of altruistic values related to the value of life, see Jones-Lee (1992).

[57] For a review of these studies, and this issue in particular see, for example, Hammitt (2007), Aldy and Viscusi (2007), and Krupnick (2007).

[58] Note that we are here reporting the total risk change that results from changes in 2020 exposures. As outlined below, this risk is not immediate - instead we model this risk as occurring with latency over the course of the ensuing 20 years.

[59] See W. Kip Viscusi, 1992, *Fatal Tradeoffs*, (Oxford University Press: New York), Table 4-1.

[60] This issue is discussed to some extent in Thomas J. Kniesner, W. Kip Viscusi, and James P. Ziliak (2010), "Policy relevant heterogeneity in the value of statistical life: New evidence from panel data quantile regressions," *Journal of Risk and Uncertainty* 40:15–31.

[61] See discussion papers provided in support of a recent EPA risk valuation workshop at http://www.epa.gov/air/toxicair/2009workshop.html (accessed November 24, 2010) in particular the papers and presentations by W. Kip Viscusi.

[62] Note that we do not employ a cessation lag for ozone mortality due to our reliance on short-term studies to estimate these benefits.

[63] The distribution of incidence results represent the uncertainty associated with the coefficient of the C-R function for each health endpoint. The distribution around the monetized benefits estimate reflects both uncertainty in the incidence as well as uncertainty associated with the valuation estimate.

[64] For a detailed description of the model, see the related report, *Uncertainty Analyses to Support the Second Section 812 Benefit-Cost Analysis of the Clean Air Act*, March 2010, and Industrial Economics, Inc. (2006).

[65] A detailed report of the case study methodology and results was completed by Industrial Economics, Inc (IEc, 2009). This report can be downloaded from the following website: www.epa.gov/oar/sect812

[66] This study is also the basis for the Inhalation Unit Risk (IUR) published on EPA's Integrated Risk Information System (IRIS) (USEPA, 1998).

[67] This estimate was based on a value presented in EPA's Cost of Illness Handbook (USEPA, 1999) for a "typical" cancer case.

[68] The chronic bronchitis value is the same as that used in EPA's Regulatory Impact Analysis (RIA) for the PM National Ambient Air Quality Standards (NAAQS) (USEPA, 2006). The non-fatal lymphoma value was derived by using the risk-risk ratio from Magat et al. (1996) along with our primary VSL estimate.

[69] Science Advisory Board (2004). *Advisory on Plans for Health Effects Analysis in the Analytical Plan for EPA's Second Prospective Analysis—Benefits and Costs of the Clean Air Act, 1990-2020: Advisory by the Health*

Effects Subcommittee of the Advisory Council on Clean Air Compliance Analysis. EPA-SAB-COUNCIL-ADV-04-002.

70 As noted earlier, a key recommendation of NRC (2008) was that ozone mortality estimates from available epidemiological studies represent a separate and additive effect to those from PM/mortality epidemiological studies.

71 See The Benefits and Costs of the Clean Air Act 1990 to 2010, USEPA Office of Air and Radiation and Office of Policy, EPA- 410-R-99-001, November 1999.

72 Alternative discount rates of three and seven percent are recommended in U.S. Environmental Protection Agency (2000). *Guidelines for Preparing Economic Analyses*, EPA 240-R-00-003, September.

73 If the quantitative magnitude of the assumption's effect on the net benefits cannot be assessed, the Project Team indicates that this is "Unknown."

74 Acid Neutralizing Capacity (ANC) is a measure of overall buffering capacity of a solution or surface waterbody. A well- buffered system will resist rapid changes in pH, while a poorly buffered system responds quickly to changes in pH. Reductions in ANC put waterbodies at risk of acidification due to this inability to buffer excess H^+ ions.

75 The CMAQ tool is described in more detail in Chapter 4 of this document.

76 U.S. Environmental Protection Agency (EPA). October 2003. *Response of surface water chemistry to the Clean Air Act Amendments of 1990.* EPA 620/R-03/001.

77 Acid deposition is calculated using the hydrogen deposition derived from both sulfur and nitrogen deposition as described in: U.S. Department of Agriculture, Forest Service, Rocky Mountain Region. January 2000. Screening Methodology for Calculating ANC Change to High Elevation Lakes: User's Guide. The deposition estimates in Figures 6-2 and 6-3 include combined wet and dry deposition for the stated years as estimated by the CMAQ modeling system version 4.6. These modeled estimates are not calibrated with monitored deposition data such as the National Atmospheric Deposition Program (NADP) data

78 Value bins for nitrogen deposition taken from: Rea, A., J. Lynch, R. White, G. Tennant, J. Phelan and N. Possiel. 2009. Critical Loads as a Policy Tool: Highlights of the NOx/SOx Secondary National Ambient Air Quality Standard Review. Slide 6: Nationwide Total Reactive Nitrogen Deposition (2002). Available online at: http://nadp.sws.uiuc.edu/meetings/fall2009/post/session4.html.

79 Percentiles are calculated using the combined nitrogen deposition data for all years and scenarios presented in the map.

80 Within the California hot spot, the modeled CMAQ ozone concentration estimates were low compared to the ozone monitoring data. This may have resulted in the eVNA analysis overestimating future ozone concentrations. This overestimate is expected to have occurred in this region for both the *with-CAAA* and *without-CAAA* scenarios, however, and therefore the effect on the difference in ozone concentrations between the two scenarios is uncertain.

81 See, for example, E.H. Lee and W.E. Hogsett. 1996. Methodology for Calculating Inputs for Ozone Secondary Standard Benefits Analysis: Part II. Prepared for the U.S. EPA, Office of Air Quality Planning and Standards, Air Quality Strategies and Standards Division. The application of laboratory-derived functions is less preferable than functions developed from field studies. However, the laboratory-derived functions frequently provide the best available information regarding the relationship between ozone exposure and crop or tree growth. The exposure-response functions applied in this report have been used in other EPA studies, such as: USEPA. July 2007. Review of the National Ambient Air Quality Standards for Ozone: Policy Assessment of Scientific and Technical information. EPA-452/R-07-007.

82 Relative yield losses are estimated instead of relative yield gains because the baseline (with CAAA) scenario in this analysis defines current conditions, whereas the counterfactual (no CAAA) scenario defines a change in current conditions. The models applied in this analysis forecast changes in yield relative to current conditions (i.e., relative to the baseline scenario).

83 Note that we performed two runs of the FASOM model, one where the response to ozone for those crop/region combinations without specific individual concentration-response functions are assumed to be zero, and a second where impacts on crop/region combinations without specific concentration-response functions were set to the values used in adjacent regions and/or proxy crops where possible (for example, soft white wheat was used for barley and sugarbeets; tomatoes for processing were used for potatoes; soybeans for fresh tomatoes; corn for fresh tomatoes if there is not a value for soybeans; etc.). We found that the difference in the overall national results between these two runs was negligible, however. As a result, in this chapter we report the results from the run that applies proxy crop/region concentration- response functions. Note further that the version of FASOM used for this analysis is the version current as of July 21, 2010.

84 See, for example, a review commissioned by USEPA for its application of FASOM to support regulatory analysis of renewable fuels standards, concluded in July of 2010 and available at the following web site (accessed November 26, 2010): http://www.epa.gov/otaq/fuels/renewablefuels/regulations

85 Perfect foresight is a basic assumption of the modeling approach on which FASOM is based. Structuring the model based on perfect foresight rather than a myopic (recursive) approach allows an expanded array of policy simulations and potential insights, which is the main purpose of this type of model.

[86] Note that the year 2000 in FASOM represents average annual activity over the 5-year period from 2000 to 2004; 2010 represents 2010 through 2014; and 2020 represents 2020 through 2024. Values provided for ozone impacts in 2000, 2010, and 2020 were applied to the 2000, 2010, and 2020 model periods in FASOM, respectively. The results presented here do not includes losses Canada and the rest of the world; for example, in 2020, higher US prices in the without-CAAA scenario result in additional consumer surplus losses to non-US consumers of $1.7 billion in the forest sector and $3.3 billion in the agricultural sector.

[87] The data was aggregated from the 36-km grid-cell level to the county level using the BenMAP version 3.0.15 "Air Quality Grid Aggregation" algorithm. The fourth quarter data is corrected for a missing day (the CMAQ runs modeled 364 days, omitting December 31) by reweighting the mean to account for the missing day.

[88] Council review of early drafts of the First Prospective analysis noted that the McClelland et al. (1991) study may not incorporate two potentially important adjustments. First, their study does not account for the "warm glow" effect, in which respondents may provide higher willingness to pay estimates simply because they favor "good causes" such as environmental improvement. Second, while the study accounts for non-response bias, it may not employ the best available methods. As a result of these concerns, a prior Council recommended that residential visibility be omitted from the overall primary benefits estimate in the First Prospective.

[89] The geographic proximity assignment is preserved for the Los Angeles and Riverside MSAs although these MSAs meet the elevation range threshold of 1500 meters. The assignment is preserved because Los Angeles is one of the study cities and also because Los Angeles has a particular set of location-specific characteristics that set it apart from Denver. As a conservative measure, Riverside MSA is also assigned to the Los Angeles study area because a significant portion of Riverside County itself is located in the South Coast Air Quality Management District, and therefore is considered by at least some measures to be part of the same regulated airshed as Los Angeles.

[90] The WTP value for San Francisco is higher than Denver, but the San Francisco value is not applied to other MSA's.

[91] The Project Team considered using the CMAQ SO_2 air quality results directly, but the decision to implement the materials damage approach described here came too late to cost-effectively recover the relevant ambient SO_2 estimates from the original CMAQ runs. The overall magnitude of the monetizable materials damage benefits is such a small part of the overall benefits of the CAAA that the impact of using APEEP's air quality tool rather than CMAQ on the overall benefits estimates is likely to be very small.

[92] Driscoll, Charles T. et al. May 2003. Chemical Response of Lakes in the Adirondack Region of New York to Declines in Acidic Deposition. Environmental Science and Technology 37(10): 2036-2042.

[93] The U.S. National Acid Precipitation Assessment Program. 1991. Integrated Assessment Report. The NAPAP Office of the Director, Washington, DC.

[94] Driscoll, Charles T. et. al. March 2001. Acidic Deposition in the Northeastern United States: Sources and Inputs, Ecosystem Effects, and Management Strategies. BioScience 51(3): 180-198.

[95] While the timeframe for ths Second Prospecti v anal sis of the CAAA is through 2020 this case tudy reports benefits through 2050 s we expect that reduct on in emis ion that occur in 2020 wil continue to provide benefits to recreationa fishing throuh this time frame.

[96] The differences between percent base saturation levels with the CAAA and without the CAAA are presented rather than absolute percent base saturation levels for each scenario to highlight the changes in percent base saturation attributable to the implementation of the CAAA.

[97] The ten ranges of difference in percent base saturation values presented in Exhibit 5-8 are equal to the 10th, 20th, …, and 100th percentiles for the combined distribution of difference in percent base saturation values across all years in the analysis (2000, 2010, 2020, and 2050).

[98] For example, Duchesne et al. (2002) found that sugar maple basal area growth rates were reduced by 17 percent, on average, in forest stands exhibiting decreasing basal area growth rates over time (declining stands) compared to sugar maple basal area growth rates in stands exhibiting increasing basal area growth rates over time (healthy stands). In addition McLaughlin (1998) found that the health of hardwood stands on shallow, poorly buffered soils similar to those found in Adirondack Park declined during the 1990s due to decreasing pH and base saturation levels and increased aluminum ion concentrations.

[99] We estimated stumpage values of commonly harvested species in the Adirondack Region by applying average stumpage values to the pulpwood and wood chip and roundwood log harvest volume estimates. The average stumpage value for pulpwood and wood chips is estimated to be $3 per ton; while, the average stumpage value for roundwood logs is estimated to be $150 per thousand board-feet (MBF). Using these estimates, the annual harvest value of pulpwood and wood chips is estimated to be approximately $5.4 million, and the annual harvest value of roundwood logs is estimated to be $15 million.

[100] Note that the direct costs were aggregated in Chapter 3.

[101] The characterization of the uncertainty surrounding economic vaLuation is discussed in detaiL in IndustriaL Economics, Inc *Uncertainty Analyses to Sipport the cond ction 812 Prospective Benefit-Cost Analysis of the Clean Air Act: I)aft* Peport, prepared for Office of Air and Radiation, US EnvironmentaL Protection Agency, AipriL 2010.

[102] The statistical aggregation technique applied is commonly referred to as Monte Carlo analysis. The technique involves many re-calculations of results, using different combinations of input parameters each time. For each calculation, values from each input parameter's statistical distribution are selected at random to ensure that the calculation does not always result in extreme values, or rely solely on low end or solely on high end input parameters. The aggregate distribution more accurately reflects a reasonable likelihood of the joint occurrence of multiple input parameters.

[103] See National Research Council (2002), *Estimating the Public Health Benefits of Proposed Air Pollution Regulations*, The National Academies Press, Washington, DC, in particular Chapter 5, titled: "Uncertainty."

[104] For a more thorough description of the methods and results of these uncertainty analyses see the accompanying *report Uncertainty Analyses to Support the Second Section 812 Benefit-Cost Analysis of the Clean Air Act*, March 2009.

[105] The estimate of the impact on total costs is derived from the relative contribution of the affected cost sector to the overall costs of compliance, assuming all other sectors are unaffected.

[106] The six were: improved emissions inventories and inventory management tools (see Chapter 2 for a description of the improvements in the 2002 NEI, and the AirControlNET tool used to estimate emissions reductions necessary for NAAQS compliance); improved tools for assessing the full range of social costs associated with regulation, including the tax-interaction effect (see Chapter 8 of this document for a description of the economic modeling tool EMPAX-CGE); a more geographically comprehensive air quality monitoring network, particularly for fine particulate matter (see discussion of the MATS procedure in Chapter 4 of this document); development of integrated air quality modeling tools based on an open, consistent model architecture (see Chapter 4 for a description of the CMAQ system); increased basic and targeted research on the health effects of air pollution, especially particulate matter (see Chapter 5 for multiple examples of recent work that was applied in this analysis); continued efforts to assess the cancer and noncancer effects of air toxics exposure (see discussion of the air toxics case study in Chapter 5).

[107] The description of CGE models, in general, and EMPAX-CGE included in this section is based on RTI (2008).

[108] Economic data and information on non-electricity energy markets are generally available only at the state level, which necessitates an approximation of the NERC regions that follows state boundaries.

[109] As noted in Chapter 2, the emissions projections for the Second Prospective are based on the economic forecast from *Annual Energy Outlook (AEO) 2005*, not *AEO 2007*. The *AEO 2007* forecast, however, is similar to that in *AEO 2005*. For the year 2020, the *AEO 2007* GDP forecast is approximately 3 percent lower than the projection from *AEO 2005*.

[110] For example, Cropper and Krupnick (1999) estimate income losses resulting from chronic bronchitis and acute myocardial infarction. Based on these estimates, we calculated the lost work time per case associated with each of these endpoints.

[111] Ibbotson Associates, *Cost of Capital Yearbook*, 1997 through 2006 editions.

[112] In most of EPA's recent EMPAX applications to air pollution rules, only a small portion of total costs have been accounted for by expenditures in the household sector. In this application, however, a large portion of total compliance costs, particularly for mobile source fuels rules, involve increased expenditures by the household sector. For this reason, the Project Team gave special consideration to the treatment of these costs. Estimated household compliance expenditures associated with petroleum products are implemented as price adjustments to reflect higher motor vehicle fuel prices. The petroleum price adjustment is calculated to match compliance expenditures related to household transportation fuel use. For other transportation compliance expenditures, the household utility function is adjusted to require additional expenditures to achieve a given utility level. These adjustments reflect the additional automotive inspections, maintenance, and technologies purchased by households to comply with the Clean Air Act. Other unidentified household compliance costs not related to transportation (e.g. non-road related local controls) are treated as lump-sum reductions to household income.

[113] For a detailed description of the model, see the related report, *Uncertainty Analyses to Support the Second Section 812 Benefit-Cost Analysis of the Clean Air Act*, March 2010, and Industrial Economics, Inc. (2006).

[114] This estimate is consistent with that used in Jorgenson et al. (2004).

[115] The EMPAX model results do not isolate the impact of the labor force effect on GDP or the impact of changes in medical expenditures, as the two were modeled simultaneously.

[116] Note that results for the labor force-adjusted case for years after 2020 indicate that the beneficial effects on the economy grow over time, through 2030, from $5 billion in 2020 to $14 billion in 2025 to $24 billion in 2030. EMPAX results for 2030, however, are considered less reliable because of the greater uncertainty in forecasting GDP and industry-level productivity 20 years into the future.

[117] We are grateful to the SAB Council for sharing this observation. For further information, see, for example, J.C. Carbone and V.K. Smith. 2008. Evaluating policy interventions with general equilibrium externalities. J. Public Econ. 92:1254-1274.

In: Better Air: Benefits and Costs of the Clean Air Act
Editor: Jessica Lincoln-Oswalt
ISBN: 978-1-61470-724-0

Chapter 2

CLEAN AIR ACT: A SUMMARY OF THE ACT AND ITS MAJOR REQUIREMENTS

James E. McCarthy, Claudia Copeland, Larry Parker and Linda-Jo Schierow

SUMMARY

This report summarizes the Clean Air Act and its major regulatory requirements.

The principal statute addressing air quality concerns, the Clean Air Act was first enacted in 1955, with major revisions in 1970, 1977, and 1990. The Act requires EPA to set health-based standards for ambient air quality, sets deadlines for the achievement of those standards by state and local governments, and requires EPA to set national emission standards for large or ubiquitous sources of air pollution, including motor vehicles, power plants, and other industrial sources. In addition, the Act mandates emission controls for sources of 187 hazardous air pollutants, establishes a capand-trade program to limit acid rain, requires the prevention of significant deterioration of air quality in areas with clean air, requires a program to restore visibility impaired by regional haze in national parks and wilderness areas, and implements the Montreal Protocol to phase out most ozone-depleting chemicals.

This report describes the Act's major provisions and provides tables listing all major amendments, with the year of enactment and Public Law number, and cross-referencing sections of the Act with the major U.S. Code sections of the codified statute.

INTRODUCTION

The authorities and responsibilities of the Environmental Protection Agency (EPA) derive primarily from a dozen major environmental statutes. This report provides a concise summary of one of those statutes, the Clean Air Act. It provides a very brief history of federal involvement in air quality regulation and of the provisions added by legislation in 1970, 1977, and 1990; it explains major authorities contained in the Act; it defines key terms; and it lists references for more detailed information on the Act and its implementation.

While this report attempts to present the essence of the Act, it is necessarily incomplete. Many details and secondary provisions are omitted. In addition, the report describes the statute largely without discussing its implementation. Statutory deadlines to control emissions and achieve particular mandates have often been missed as a result of delayed standard-setting by EPA, delayed action on implementation by states and local governments, or law suits brought by interested parties.

OVERVIEW

The Clean Air Act, codified as 42 U.S.C. 7401 *et seq*., seeks to protect human health and the environment from emissions that pollute ambient, or outdoor, air. It requires the Environmental Protection Agency to establish minimum national standards for air quality, and assigns primary responsibility to the states to assure compliance with the standards. Areas not meeting the standards, referred to as "nonattainment areas," are required to implement specified air pollution control measures. The Act establishes federal standards for mobile sources of air pollution and their fuels and for sources of 187 hazardous air pollutants, and it establishes a cap-and-trade program for the emissions that cause acid rain. It establishes a comprehensive permit system for all major sources of air pollution. It also addresses the prevention of pollution in areas with clean air and protection of the stratospheric ozone layer.

Like many other programs administered by the Environmental Protection Agency, federal efforts to control air pollution have gone through several phases, beginning with information collection, research, and technical assistance, before being strengthened to establish federal standards and enforcement. Federal legislation addressing air pollution was first passed in 1955, prior to which air pollution was the exclusive responsibility of state and local levels of government.

The federal role was strengthened in subsequent amendments, notably the Clean Air Act Amendments of 1970, 1977, and 1990. The 1970 amendments established the procedures under which EPA sets national standards for air quality, required a 90% reduction in emissions from new automobiles by 1975, established a program to require the best available control technology at major new sources of air pollution, established a program to regulate air toxics, and greatly strengthened federal enforcement authority. The 1977 amendments adjusted the auto emission standards, extended deadlines for the attainment of air quality standards, and added the Prevention of Significant Deterioration program to protect air cleaner than national standards.

Changes to the Act in 1990 included provisions to (1) classify most nonattainment areas according to the extent to which they exceed the standard, tailoring deadlines, planning, and controls to each area's status; (2) tighten auto and other mobile source emission standards; (3) require reformulated and alternative fuels in the most polluted areas; (4) revise the air toxics section, establishing a new program of technology-based standards and addressing the problem of sudden, catastrophic releases of air toxics; (5) establish an acid rain control program, with a marketable allowance scheme to provide flexibility in implementation; (6) require a state-run permit program for the operation of major sources of air pollutants; (7)

implement the Montreal Protocol to phase out most ozone-depleting chemicals; and (8) update the enforcement provisions so that they parallel those in other pollution control acts, including authority for EPA to assess administrative penalties.

The 1990 amendments also authorized appropriations for clean air programs through FY1998. The Act has not been reauthorized since then. House rules require enactment of an authorization before an appropriation bill can be considered; but this requirement can be waived and frequently has been. Thus, while authorization of appropriations in the Clean Air Act (and most other environmental statutes) has expired, programs have continued and have been funded. The Act's other legal authorities, to issue and enforce regulations, are, for the most part, permanent and are not affected by the lack of authorization.

The remainder of this report describes major programs required by the Act, with an emphasis on the changes established by the 1990 amendments.

Table 1. Clean Air Act and Amendments (codified generally as 42 U.S.C. 7401-7671).

Year	Act	Public Law Number
1955	Air Pollution Control Act	P.L. 84-159
1959	Reauthorization	P.L. 86-35 3
1960	Motor vehicle exhaust study	P.L. 86-493
1963	Clean Air Act Amendments	P.L. 88-206
1965	Motor Vehicle Air Pollution Control Act	P.L. 89-272, Title I
1966	Clean Air Act Amendments of 1966	P.L. 89-675
1967	Air Quality Act of 1967	P.L. 90-148
	National Air Emission Standards Act	
1970	Clean Air Act Amendments of 1970	P.L. 91-604
1973	Reauthorization	P.L. 93-13
1974	Energy Supply and Environmental Coordination Act of 1974	P.L. 93-319
1977	Clean Air Act Amendments of 1977	P.L. 95-95
1980	Acid Precipitation Act of 1980	P.L. 96-294, Title VII
1981	Steel Industry Compliance Extension Act of 1981	P.L. 97-23
1987	Clean Air Act 8-month Extension	P.L. 100-202
1990	Clean Air Act Amendments of 1990	P.L. 101-549
1991	Technical correction to list of hazardous air pollutants	P.L. 102-187
1995-96	Relatively minor laws amending the Act	P.L. 104-6, 59, 70, 260
1998	Amended § 604 re methyl bromide	P.L. 105-277, Section 764
1998	Border Smog Reduction Act of 1998	P.L. 105-286
1999	Chemical Safety Information, Site Security and Fuels Regulatory Relief Act	P.L. 106-40
2004	Amendments to §209 re small engines	P.L. 108-199, Division G, Title IV, Section 428
2005	Energy Policy Act of 2005 (amended §211 re fuels)	P.L. 109-58
2007	Energy Independence and Security Act of 2007 (amended §211 re fuels)	P.L. 110-140

National Ambient Air Quality Standards

In section 109, the Act requires EPA to establish National Ambient Air Quality Standards (NAAQS) for air pollutants that endanger public health or welfare, in the Administrator's judgment, and whose presence in ambient air results from numerous or diverse sources. The NAAQS must be designed to protect public health and welfare with an adequate margin of safety. Using this authority, EPA has promulgated NAAQS for six air pollutants: sulfur dioxide (SO_2), particulate matter (PM2.5 and PM10), nitrogen dioxide (NO_2), carbon monoxide (CO), ozone,[1] and lead. The Act requires EPA to review the scientific data upon which the standards are based, and revise the standards, if necessary, every five years. More often than not, however, EPA has taken more than five years in reviewing and revising the standards.

Originally, the Act required that the NAAQS be attained by 1977 at the latest, but the states experienced widespread difficulty in complying with this deadline. As a result, the deadlines for achieving NAAQS have been extended several times. Under the 1990 amendments, most areas not in attainment with NAAQS must meet special compliance schedules, staggered according to the severity of an area's air pollution problem. The amendments also established specific requirements for each nonattainment category, as described below.

State Implementation Plans

While the Act authorizes the EPA to set NAAQS, the states are responsible for establishing procedures to attain and maintain the standards. Under Section 110 of the Act, the states adopt plans, known as State Implementation Plans (SIPs), and submit them to EPA to ensure that they are adequate to meet statutory requirements.

SIPs are based on emission inventories and computer models to determine whether air quality violations will occur. If these data show that standards would be exceeded, the state must impose additional controls on existing sources to ensure that emissions do not cause "exceedances" of the standards. Proposed new and modified sources must obtain state construction permits in which the applicant shows how the anticipated emissions will not exceed allowable limits. In nonattainment areas, emissions from new or modified sources must also be offset by reductions in emissions from existing sources.

The 1990 amendments require EPA to impose sanctions in areas which fail to submit a SIP, fail to submit an adequate SIP, or fail to implement a SIP: unless the state corrects such failures, a 2-to-1 emissions offset for the construction of new polluting sources is imposed 18 months after notification to the state, and a ban on most new federal highway grants is imposed six months later. An additional ban on air quality grants is discretionary. Ultimately, a Federal Implementation Plan may be imposed if the state fails to submit or implement an adequate SIP.

The amendments also require that, in nonattainment areas, no federal permits or financial assistance may be granted for activities that do not "conform" to a State Implementation Plan. This requirement can cause a temporary suspension in funding for most new highway and transit projects if an area fails to demonstrate that the emissions caused by such projects are consistent with attainment and maintenance of ambient air quality standards. Demonstrating conformity of transportation plans and SIPs is required in nonattainment areas whenever new plans are submitted.

NONATTAINMENT REQUIREMENTS

In a major departure from the prior law, the 1990 Clean Air Act Amendments grouped most nonattainment areas into classifications based on the extent to which the NAAQS was exceeded, and established specific pollution controls and attainment dates for each classification. These requirements are described here as spelled out in Sections 181-193 of the Act.[2]

Nonattainment areas are classified on the basis of a "design value," which is derived from the pollutant concentration (in parts per million or micrograms per cubic meter) recorded by air quality monitoring devices. The design value for the 1-hour ozone standard was the fourth highest hourly reading measured during the most recent three-year period. Using these design values, the Act created five classes of ozone nonattainment, as shown in **Table 2**. Only Los Angeles fell into the "extreme" class, but 97 other areas were classified in one of the other four ozone categories. A simpler classification system established moderate and serious nonattainment areas for carbon monoxide and particulate matter with correspondingly more stringent control requirements for the more polluted class.

As shown in the table, the deadlines for attainment for ozone nonattainment areas stretched from 1993 to 2010, depending on the severity of the problem. (Under the 8-hour ozone standard, which replaced the 1-hour standard in 2004, these deadlines are changed to 2007 to 2021.) For carbon monoxide, the attainment date for moderate areas was December 31, 1995, and for serious areas, December 31, 2000. For particulate matter, the deadline for areas designated moderate nonattainment as of 1990 was December 31, 1994; for those areas subsequently designated as moderate, the deadline is six years after designation. For serious areas, the respective deadlines are December 31, 2001 or 10 years after designation.

Table 2. Ozone Nonattainment Classifications.

Class	Marginal	Moderate	Serious	Severe	Extreme
Deadline	1993	1996	1999	2005-2007[a]	2010
Areas[b]	42 areas	32 areas	14 areas	9 areas	1 area
Design Value	0.121 ppm- 0.138 ppm	0.138 ppm- 0.160 ppm	0.160 ppm- 0.180 ppm	0.180 ppm- 0.280 ppm	>0.280 ppm

a. Areas with a 1988 design value between 0.190 and 0.280 ppm have 17 years to attain; others have 15 years.

b. Number of areas in each category as of the date of enactment.

Requirements for Ozone Nonattainment Areas

Although areas with more severe air pollution problems have a longer time to meet the standards, more stringent control requirements are imposed in areas with worse pollution. A summary of the primary ozone control requirements for each nonattainment category follows.

Marginal Areas

- Inventory emissions sources (to be updated every three years).
- Require 1.1 to 1 offsets (i.e., new major emission sources of volatile organic compounds [VOCs] must reduce VOC emissions from existing facilities in the area by 10% more than the emissions of the new facility).
- Impose reasonably available control technology (RACT) on all major sources emitting more than 100 tons per year for the nine industrial categories where EPA had already issued control technique guidelines describing RACT prior to 1990.

Moderate Areas

- Meet all requirements for marginal areas.
- Impose a 15% reduction in VOC emissions in six years.
- Adopt a basic vehicle inspection and maintenance program.
- Impose RACT on all major sources emitting more than 100 tons per year for all additional industrial categories where EPA will issue control technique guidelines describing RACT.
- Require vapor recovery at gas stations selling more than 10,000 gallons per month.
- Require 1.15 to 1 offsets.

Serious Areas

- Meet all requirements for moderate areas.
- Reduce definition of a major source of VOCs from emissions of 100 tons per year to 50 tons per year for the purpose of imposing RACT.
- Reduce VOCs 3% annually for years 7 to 9 after the 15% reduction already required by year 6.
- Improve monitoring.
- Adopt an enhanced vehicle inspection and maintenance program.
- Require fleet vehicles to use clean alternative fuels.
- Adopt transportation control measures if the number of vehicle miles traveled in the area is greater than expected.
- Require 1.2 to 1 offsets.
- Adopt contingency measures if the area does not meet required VOC reductions.

Severe Areas

- Meet all requirements for serious areas.
- Reduce definition of a major source of VOCs from emissions of 50 tons per year to 25 tons per year for the purpose of imposing RACT.
- Adopt specified transportation control measures.
- Implement a reformulated gasoline program.
- Require 1.3 to 1 offsets.
- Impose $5,000 per ton penalties on major sources if the area does not meet required reductions.

Extreme Areas

- Meet all requirements for severe areas.
- Reduce definition of a major source of VOCs from emissions of 25 tons per year to 10 tons per year for the purpose of imposing RACT.
- Require clean fuels or advanced control technology for boilers emitting more than 25 tons per year of NO_x.
- Require 1.5 to 1 offsets.

As noted, EPA promulgated a new, 8-hour ozone standard in July 1997. Following extensive court challenges, the agency designated nonattainment areas for the new standard on April 30, 2004. State Implementation Plans were required to be submitted in 2007.

Requirements for Carbon Monoxide Nonattainment Areas

As with ozone nonattainment areas, carbon monoxide (CO) nonattainment areas are subjected to specified control requirements, with more stringent requirements in Serious nonattainment areas. A summary of the primary CO control requirements for each nonattainment category follows.

Moderate Areas

- Conduct an inventory of emissions sources.
- Forecast total vehicle miles traveled in the area.
- Adopt an enhanced vehicle inspection and maintenance program.
- Demonstrate annual improvements sufficient to attain the standard.

Serious Areas

- Adopt specified transportation control measures.
- Implement an oxygenated fuels program for all vehicles in the area.
- Reduce definition of a major source of CO from emissions of 100 tons per year to 50 tons per year if stationary sources contribute significantly to the CO problem.

Serious areas failing to attain the standard by the deadline have to revise their SIP and demonstrate reductions of 5% per year until the standard is attained.

Requirements for Particulate Nonattainment Areas

Particulate (PM10) nonattainment areas are also subject to specified control requirements. These are:

Moderate Areas

- Require permits for new and modified major stationary sources of PM10.
- Impose reasonably available control measures (RACM).

Serious Areas

- Impose best available control measures (BACM).
- Reduce definition of a major source of PM10 from 100 tons per year to 70 tons per year.

In July 1997, EPA promulgated new standards for fine particulates (PM2.5). The PM2.5 standards were also subject to court challenges. The absence of a monitoring network capable of measuring the pollutant delayed implementation as well. Nonattainment areas for PM2.5 were designated on April 14, 2005. States had three years subsequent to designation to submit State Implementation Plans. Revisions to the NAAQS promulgated in October 2006 strengthened the PM2.5 standard.

TRANSPORTED AIR POLLUTION

Meeting the nation's clean air standards can be complicated as air pollution is no respecter of political boundaries or subdivisions. This problem of transported air pollutants has come into particular focus as states and EPA attempt to develop effective compliance strategies to achieve both the ozone and the PM2.5 NAAQS. Under Section 110(a)(2)(D), SIPs must include adequate provisions to prevent sources within that state from contributing significantly to nonattainment in one or more downwind states.

If EPA finds a SIP inadequate to achieve a NAAQS, it must require the affected state to submit a revised SIP that includes sufficient measures to bring that state into compliance. This is known as a "SIP Call." The 1990 Clean Air Act amendments provided EPA and the states with new tools to address the transport problem through this provision. One of those tools is Section 176A, a provision that permits the EPA, either on its own or by petition from any state, to establish a transport region to address regional pollution problems contributing to violations of a primary NAAQS. A commission of EPA and state officials is constituted to make recommendations to EPA on appropriate mitigation strategies. Based on the commission's findings and recommendations, EPA is then required under section 1 10(k)(5) to notify affected states of inadequacies in their current state implementation plans and to establish deadlines (not to exceed 18 months) for submitting necessary revisions (i.e., a SIP call). Besides authorizing administratively-created transport regions, the 1990 amendments statutorily created an Ozone Transport Region (OTR) in the Northeast. This provision (Section 184 of the act) required specific additional controls for all areas (not only nonattainment areas) in that region, and established the Ozone Transport Commission for the purpose of recommending to EPA regionwide controls affecting all areas in the region.

The transport issue may also be addressed by affected downwind states through a Section 126 petition. As amended by the 1990 Clean Air Act amendments, under Section 126(b) any state or political subdivision may petition EPA for a finding that a major source or group of stationary sources located in another state is emitting pollutants that "significantly contribute" to the nonattainment of a NAAQS by their state. EPA is to respond to the petition within 60 days. If the petition is granted, the offending sources must cease operations within three months unless the sources comply with emission controls and the compliance schedules as determined by EPA to bring them into compliance with the section. Section 126 has rarely been used, although it has proven useful to EPA in some cases as backup authority where there might be challenges to a SIP call.

Emission Standards for Mobile Sources

Title II of the Clean Air Act has required emission standards for automobiles since 1968. The 1990 amendments significantly tightened these standards: for cars, the hydrocarbon standard was reduced by 40% and the nitrogen oxides (NO_x) standard by 50%. These standards—referred to as "Tier 1" standards—were phased in over the 1994-1996 model years.

The amendments envisioned a further set of reductions ("Tier 2" standards), but not before model year 2004. For Tier 2 standards to be promulgated, the agency was first required to report to Congress concerning the need for further emission reductions, the availability of technology to achieve such reductions, and the cost-effectiveness of such controls compared to other means of attaining air quality standards. EPA submitted this report to Congress in August 1998, concluding that further emission reductions were needed and that technology to achieve such reductions was available and cost-effective. Tier 2 standards, requiring emission reductions of 77% to 95% from cars and light trucks were promulgated in February 2000, and were phased in over the 2004-2009 model years. To facilitate the use of more effective emission controls, the standards also require a more than 90% reduction in the sulfur content of gasoline, beginning in 2004.

The 1990 amendments also required that oxygenated gasoline, designed to reduce emissions of carbon monoxide, be sold in the worst CO nonattainment areas and that "reformulated" gasoline (RFG), designed to reduce emissions of volatile organic compounds and toxic air pollutants, be sold in the nine worst ozone nonattainment areas (Los Angeles, San Diego, Houston, Baltimore, Philadelphia, New York, Hartford, Chicago, and Milwaukee); metropolitan Washington, DC; and four areas in California were added to the mandatory list later. Other ozone nonattainment areas can opt in to the RFG program; as of 2006, additional areas in 11 states had done so.

The fuels provisions were modified by the Energy Policy Act of 2005 (EPACT), removing the requirement that RFG contain oxygenates. Instead, EPACT required the use of increasing amounts of renewable fuel, most likely to be ethanol, in motor fuels, beginning in 2006. The Energy Independence and Security Act of 2007 further strengthened the renewable fuel requirements.

Use of alternative fuels and development of cleaner engines was also to be stimulated by the Clean-Fuel Fleet Program. In all of the most seriously polluted ozone and CO nonattainment areas, centrally fueled fleets of 10 or more passenger cars and light-duty trucks must purchase at least 30% clean-fuel vehicles when they add new vehicles to existing fleets, starting in 1999. (The Act originally required the program to begin in 1998, but the start was delayed by a year.) The percentage rose to 50% in 2000 and 70% in 2001. Heavy-duty fleets are required to purchase at least 50% clean-fuel vehicles annually. A clean fuel vehicle is one which meets Low Emission Vehicle (LEV) standards and operates on reformulated gasoline, reformulated diesel, methanol, ethanol, natural gas, liquefied petroleum gas, hydrogen, or electricity.

In addition to the above program, California's Zero Emission Vehicle (ZEV) program also is intended to promote the development of alternative fuels and vehicles. Section 209(b) of the Clean Air Act allows the EPA Administrator to grant California the authority to develop its own vehicle emissions standards if those standards are at least as stringent as the federal standards and if the state demonstrates that it needs the standards to meet compelling and extraordinary conditions. In addition to setting more stringent standards for all vehicles, California used this authority to establish a program requiring auto manufacturers to sell ZEVs (electric or hydrogen fuel cell vehicles) in the state beginning in 2003. This program has been substantially modified since it was enacted, and now allows credit for hybrid and partial ZEV vehicles in addition to true ZEVs, but it has served as an incubator for lower emission technologies since its adoption. Section 177 of the Act allows other states to adopt California's stricter standards: at least ten states (Connecticut, Maine, Massachusetts, New Jersey, New York, Oregon, Pennsylvania, Rhode Island, Vermont, and Washington) have already adopted them or are in the process of doing so.

The 1990 amendments also imposed tighter requirements on certification (an auto's useful life is defined as 100,000 miles instead of the earlier 50,000 miles), on emissions allowed during refueling, on low temperature CO emissions, on in-use performance over time, and on warranties for the most expensive emission control components (8 years/80,000 miles for the catalytic converter, electronic emissions control unit, and onboard emissions diagnostic unit). Regulations were also extended to include nonroad fuels and engines.

Standards for trucks and buses using diesel engines were also strengthened. The 1990 amendments required new urban buses to reduce emissions of diesel particulates 92% by 1996, and all other heavy-duty diesel engines to achieve an 83% reduction by the same year. NO_x emissions must also be reduced, 33% by 1998. Authority to further strengthen these standards led to promulgation in January 2001 of new emission standards requiring a further 90%-95% reduction in emissions phased in over the 2007-2010 model years, and a reduction of 97% in the allowable amount of sulfur in highway diesel fuel. These regulations were followed in May 2004 by similar requirements for nonroad diesel equipment, which will be phased in between 2007 and 2015.

HAZARDOUS AIR POLLUTANTS[3]

Completely rewritten by the Clean Air Act Amendments of 1990, Section 112 of the Act establishes programs for protecting public health and the environment from exposure to toxic air pollutants. As revised by the 1990 amendments, the section contains four major provisions: Maximum Achievable Control Technology (MACT) requirements; health-based standards; standards for stationary "area sources" (small, but numerous sources, such as gas stations or dry cleaners, that collectively emit significant quantities of hazardous pollutants); and requirements for the prevention of catastrophic releases.

First, EPA is to establish technology-based emission standards, called MACT standards, for sources of 187 pollutants listed in the legislation, and to specify categories of sources subject to the emission standards.[4] EPA is to revise the standards periodically (at least every eight years). EPA can, on its own initiative or in response to a petition, add or delete substances or source categories from the lists.

Section 112 establishes a presumption in favor of regulation for the designated chemicals; it requires regulation of a designated pollutant unless EPA or a petitioner is able to show "that there is adequate data on the health and environmental effects of the substance to determine that emissions, ambient concentrations, bioaccumulation or deposition of the substance may not reasonably be anticipated to cause any adverse effects to human health or adverse environmental effects."

EPA is required to set standards for sources of the listed pollutants that achieve "the maximum degree of reduction in emissions" taking into account cost and other non-air-quality factors. These MACT standards for new sources "shall not be less stringent than the most stringent emissions level that is achieved in practice by the best controlled similar source." The standards for existing sources may be less stringent than those for new sources, but must be no less stringent than the emission limitations achieved by either the best performing 12% of existing sources (if there are more than 30 such sources in the category or subcategory) or the best performing 5 similar sources (if there are fewer than 30). Existing sources are given three years following promulgation of standards to achieve compliance, with a possible one-year extension; additional extensions may be available for special circumstances or for certain categories of sources. Existing sources that achieve voluntary early emissions reductions will receive a six-year extension for compliance with MACT.

The second major provision of Section 112 directs EPA to set health-based standards to address situations in which a significant residual risk of adverse health effects or a threat of adverse environmental effects remains after installation of MACT. This provision requires that EPA, after consultation with the Surgeon General of the United States, submit a report to Congress on the public health significance of residual risks, and recommend legislation regarding such risks. If Congress does not legislate in response to EPA's recommendations, then EPA is required to issue standards for categories of sources of hazardous air pollutants as necessary to protect the public health with an ample margin of safety or to prevent an adverse environmental effect. A residual risk standard is required for any source emitting a cancer-causing pollutant that poses an added risk to the most exposed person of more than one-in-a-million. Residual risk standards are due eight years after promulgation of MACT for the affected source category. Existing sources have 90 days to comply with a residual risk standard, with a possible two-year extension. In general, residual risk standards do not apply to area sources.

The law directed EPA to contract with the National Academy of Sciences (NAS) for a study of risk assessment methodology, and created a Risk Assessment and Management Commission to investigate and report on policy implications and appropriate uses of risk assessment and risk management. In 1994 NAS published its report, *Science and Judgment in Risk Assessment.* The Commission study, *Framework for Environmental Health Risk Management*, was released in 1997.

Third, in addition to the technology-based and health-based programs for major sources of hazardous air pollution, EPA is to establish standards for stationary "area sources" determined to present a threat of adverse effects to human health or the environment. The provision requires EPA to regulate the stationary area sources responsible for 90% of the emissions of the 30 hazardous air pollutants that present the greatest risk to public health in the largest number of urban areas. In setting the standard, EPA can impose less stringent "generally available" control technologies, rather than MACT.

Finally, Section 112 addresses prevention of sudden, catastrophic releases of air toxics by establishing an independent Chemical Safety and Hazard Investigation Board. The Board is responsible for investigating accidents involving releases of hazardous substances, conducting studies, and preparing reports on the handling of toxic materials and measures to reduce the risk of accidents.

EPA is also directed to issue prevention, detection, and correction requirements for catastrophic releases of air toxics by major sources. Section 112(r) requires owners and operators to prepare risk management plans including hazard assessments, measures to prevent releases, and a response program.

NEW SOURCE PERFORMANCE STANDARDS[5]

Section 111 of the Act requires EPA to establish nationally uniform, technology-based standards (called New Source Performance Standards, or NSPS) for categories of new industrial facilities. These standards accomplish two goals: first, they establish a consistent baseline for pollution control that competing firms must meet, and thereby remove any incentive for states or communities to weaken air pollution standards in order to attract polluting industry; and second, they preserve clean air to accommodate future growth, as well as for its own benefits.

NSPS establish maximum emission levels for new major stationary sources—powerplants, steel mills, and smelters, for example—with the emission levels determined by the best "adequately demonstrated" continuous control technology available, taking costs into account. EPA must regularly revise and update NSPS applicable to designated sources as new technology becomes available, since the goal is to prevent new pollution problems from developing and to force the installation of new control technology.

The standards also apply to modifications of existing facilities, through a process called New Source Review (NSR). The law's ambiguity regarding what constitutes a modification (subject to NSR) as opposed to routine maintenance of a facility has led to litigation, with EPA proposing in recent years to modify its interpretation of the requirements of this section.

Solid Waste Incinerators

Prior to 1990, solid waste incinerators, which emit a wide range of pollutants, were subject to varying degrees of state and federal regulation depending on their size, age, and the type of waste burned. In a new Section 129, the 1990 amendments established more consistent federal requirements specifying that emissions of 10 categories of pollutants be regulated at new and existing incinerators burning municipal solid waste, medical waste, and commercial and industrial waste. The amendments also established emissions monitoring and operator training requirements.

Prevention of Significant Deterioration / Regional Haze[6]

Sections 160-169 of the act establish requirements for the prevention of significant deterioration of air quality (PSD). The PSD program reflects the principle that areas where air quality is better than that required by NAAQS should be protected from significant new air pollution even if NAAQS would not be violated.

The Act divides clean air areas into three classes, and specifies the increments of SO_2 and particulate pollution allowed in each. Class I areas include international and national parks, wilderness and other pristine areas; allowable increments of new pollution are very small. Class II areas include all attainment and not classifiable areas, not designated as Class I; allowable increments of new pollution are modest. Class III represents selected areas that states may designate for development; allowable increments of new pollution are large (but not exceeding NAAQS). Through an elaborate hearing and review process, a state can have regions redesignated from Class II to Class III (although none have yet been so redesignated).

While the 1977 amendments only stipulated PSD standards for two pollutants, SO_2 and particulates, EPA is supposed to establish standards for other criteria pollutants. Thus far, only one of the other four (NO_2) has been addressed: the agency promulgated standards for NO_2 in 1988.

Newly constructed polluting sources in PSD areas must install best available control technology (BACT) that may be more strict than that required by NSPS. The justifications of the policy are that it protects air quality, provides an added margin of health protection, preserves clean air for future development, and prevents firms from gaining a competitive edge by "shopping" for clean air to pollute.

In Sections 1 69A and B, the Act also sets a national goal of preventing and remedying impairment of visibility in national parks and wilderness areas, and requires EPA to promulgate regulations to assure reasonable progress toward that goal. In the 1990 Amendments, Congress strengthened these provisions, which had not been implemented.

The amendments required EPA to establish a Grand Canyon Visibility Transport Commission, composed of Governors from each state in the affected region, an EPA designee, and a representative of each of the national parks or wilderness areas in the region. Other visibility transport commissions can be established upon EPA's discretion or upon petition from at least two states. Within 18 months of receiving a report from one of these commissions, EPA is required to promulgate regulations to assure reasonable progress toward the visibility goal, including requirements that states update their State Implementation Plans to contain emission limits, schedules of compliance, and other measures necessary to make reasonable progress. Specifically mentioned is a requirement that states impose Best Available Retrofit Technology on existing sources of emissions impairing visibility.

The Grand Canyon Commission delivered a set of recommendations to EPA in June 1996, and the agency subsequently promulgated a "regional haze" program applicable to all 50 states under this authority.

ACID DEPOSITION CONTROL[7]

The Clean Air Act Amendments of 1990 added an acid deposition control program (Title IV) to the Act. It set goals for the year 2000 of reducing annual SO_2 emissions by 10 million tons from 1980 levels and reducing annual NO_x emissions by 2 million tons, also from 1980 levels.

The SO_2 reductions were imposed in two steps. Under Phase 1, owners/operators of 111 electric generating facilities listed in the law that are larger than 100 megawatts had to meet tonnage emission limitations by January 1, 1995. This would reduce SO_2 emission by about 3.5 million tons. Phase 2 included facilities larger than 75 megawatts, with a deadline of January 1, 2000. Compliance has been 100%.

To introduce some flexibility in the distribution and timing of reductions, the Act creates a comprehensive permit and emissions allowance system. An allowance is a limited authorization to emit a ton of SO_2. Issued by EPA, the allowances would be allocated to Phase 1 and Phase 2 units in accordance with baseline emissions estimates. Powerplants which commence operation after November 15, 1990 would not receive any allowances. These new units would have to obtain allowances (offsets) from holders of existing allowances. Allowances may be traded nationally during either phase. The law also permitted industrial sources and powerplants to sell allowances to utility systems under regulations developed by EPA. Allowances may be banked by a utility for future use or sale.

The Act provided for two types of sales to improve the liquidity of the allowance system and to ensure the availability of allowances for utilities and independent power producers who need them. First, a special reserve fund consisting of 2.8% of Phase 1 and Phase 2 allowance allocations has been set aside for sale. Allowances from this fund (25,000 annually from 1993-1999 and 50,000 thereafter) are sold at a fixed price of $1,500 an allowance. Independent power producers have guaranteed rights to these allowances under certain conditions. Second, an annual, open auction sold allowances (150,000 from 1993-1995, and 250,000 from 1996-1999) with no minimum price. Utilities with excess allowances may have them auctioned off at this auction, and any person may buy allowances.

The Act essentially caps SO_2 emissions at individual existing sources through a tonnage limitation, and at future plants through the allowance system. First, emissions from most existing sources are capped at a specified emission rate times an historic baseline level. Second, for plants commencing operation after November 15, 1990, emissions must be completely offset with additional reductions at existing facilities beginning after Phase 2 compliance. However, as noted above, the law provides some allowances to future powerplants which meet certain criteria. The utility SO_2 emission cap was set at 8.9 million tons, with some exceptions.

The Act provides that if an affected unit does not have sufficient allowances to cover its emissions, it is subject to an excess emission penalty of $2,000 per ton of SO_2 and required to reduce an additional ton of SO_2 the next year for each ton of excess pollutant emitted.

The Act also requires EPA to inventory industrial emissions of SO_2 and to report every five years, beginning in 1995. If the inventory shows that industrial emissions may reach levels above 5.60 million tons per year, then EPA is to take action under the Act to ensure that the 5.60 million ton cap is not exceeded.

The Act requires EPA to set specific NO_x emission rate limitations—0.45 lb. per million Btu for tangentially-fired boilers and 0.50 lb. per million Btu for wall-fired boilers—unless those rates can not be achieved by low-NO_x burner technology. Tangentially and wall-fired boilers affected by Phase 1 SO_2 controls must also meet NO_x requirements. EPA was to set emission limitations for other types of boilers by 1997 based on low-NO_x burner costs, which EPA did. In addition, EPA was to propose and promulgate a revised new source performance standard for NO_x from fossil fuel steam generating units, which EPA also did, in 1998.

PERMITS[8]

The Clean Air Act Amendments of 1990 added a Title V to the Act which requires states to administer a comprehensive permit program for the operation of sources emitting air pollutants. These requirements are modeled after similar provisions in the Clean Water Act. Previously, the Clean Air Act contained limited provision for permits, requiring only new or modified major stationary sources to obtain construction permits (under Section 165 of the Act).

Sources subject to the permit requirements generally include major sources that emit or have the potential to emit 100 tons per year of any regulated pollutant, plus stationary and area sources that emit or have potential to emit lesser specified amounts of hazardous air pollutants. However, in nonattainment areas, the permit requirements also include sources which emit as little as 50, 25, or 10 tons per year of VOCs, depending on the severity of the region's nonattainment status (serious, severe, or extreme).

States were required to develop permit programs and to submit those programs for EPA approval by November 15, 1993. EPA had one year to approve or disapprove a state's submission in whole or in part. After the effective date of a state plan, sources had 12 months to submit an actual permit application.

States are to collect annual fees from sources sufficient to cover the "reasonable costs" of administering the permit program, with revenues to be used to support the agency's air pollution control program. The fee must be at least $25 per ton of regulated pollutants (excluding carbon monoxide). Permitting authorities have discretion not to collect fees on emissions in excess of 4,000 tons per year and may collect other fee amounts, if appropriate.

The permit states how much of which air pollutants a source is allowed to emit. As a part of the permit process, a source must prepare a compliance plan and certify compliance. The term of permits is limited to no more than five years; sources are required to renew permits at that time. State permit authorities must notify contiguous states of permit applications that may affect them; the application and any comments of contiguous states must be forwarded to EPA for review. EPA can veto a permit; however, this authority is essentially limited to major permit changes. EPA review need not include permits which simply codify elements of a state's overall clean air plan, and EPA has discretion to not review permits for small sources. Holding a permit to some extent shields a source from enforcement actions: the Act provides that a source cannot be held in violation if it is complying with explicit requirements addressed in a permit, or if the state finds that certain provisions do not apply to that source.

ENFORCEMENT

Section 113 of the Act, which was also strengthened by the 1990 amendments, covers enforcement. The section establishes federal authority to issue agency and court orders requiring compliance and to impose penalties for violations of Act requirements. Section 114 authorizes EPA to require sources to submit reports, monitor emissions, and certify compliance with the Act's requirements, and authorizes EPA personnel to conduct inspections.

Like most federal environmental statutes, the Clean Air Act is enforced primarily by states or local governments; they issue most permits, monitor compliance, and conduct the majority of inspections. The federal government functions as a backstop, with authority to review state actions. The agency may act independently or may file its own enforcement action in cases where it concludes that a state's response was inadequate.

The Act also provides for citizen suits both against persons (including corporations or government agencies) alleged to have violated emissions standards or permit requirements, and against EPA in cases where the Administrator has failed to perform an action that is not discretionary under the Act. Citizen groups have often used the latter provision to compel the Administrator to promulgate regulations required by the statute.

The 1990 Amendments elevated penalties for some knowing violations from misdemeanors to felonies; removed the ability of a source to avoid an enforcement order or civil penalty by ceasing a violation within 60 days of notice; gave authority to EPA to assess administrative penalties; and authorized $10,000 awards to persons supplying information leading to convictions under the Act.

STRATOSPHERIC OZONE PROTECTION[9]

Title VI of the 1990 Clean Air Act Amendments represents the United States' primary response on the domestic front to the ozone depletion issue. It also implements the U.S. international responsibilities under the Montreal Protocol on Substances that Deplete the Ozone Layer (and its amendments). Indeed, Section 606(a)(3) provides that the Environmental Protection Agency shall adjust phase-out schedules for ozone-depleting substances in accordance with any future changes in Montreal Protocol schedules. As a result, the phase-out schedules contained in Title VI for various ozone-depleting compounds have now been superseded by subsequent amendments to the Montreal Protocol.

Since passage of Title VI, depleting substances such as CFCs, methyl chloroform, carbon tetrachloride, and halons (referred to as Class 1 substances) have been phased out by industrial countries, including the United States. New uses of hydrochlorofluorocarbons

(HCFCs) (called Class 2 substances under Title VI) are banned beginning January 1, 2015, unless the HCFCs are recycled, used as a feedstock, or used as a refrigerant for appliances manufactured prior to January 1, 2020. Production of HCFCs is to be frozen January 1, 2015 and phased out by January 1, 2030. Exemptions consistent with the Montreal Protocol are allowed.

The EPA is required to add any substance with an ozone depletion potential (ODP) of 0.2 or greater to the list of Class 1 substances and set a phase-out schedule of no more than seven years. For example, methyl bromide (ODP estimated by EPA at 0.7) was added to the list in December 1993, requiring its phaseout by January 1, 2001; this decision was altered by Congress in 1998 to harmonize the U.S. methyl bromide phase-out schedule with the 2005 deadline set by the parties to the Montreal Protocol in 1997. Also, EPA is required to add any substance that is known or may be reasonably anticipated to harm the stratosphere to the list of Class 2 substances and set a phase-out schedule of no more than ten years.

Title VI contains several implementing strategies to avoid releases of ozone-depleting chemicals to the atmosphere, including (1) for Class 1 substances used as refrigerant—lowest achievable level of use and emissions, maximum recycling, and safe disposal required by July 1, 1992; (2) for servicing or disposing refrigeration equipment containing Class 1 and 2 substances—venting banned as of July 1, 1992; (3) for motor vehicle air conditioners containing Class 1 or 2 substances—recycling required by January 1, 1992 (smaller shops by January 1, 1993); (4) sale of small containers of class 1 and 2 substances—banned within two years of enactment; and (5) nonessential products—banned within two years of enactment.

SELECTED REFERENCES

[1] U.S. Environmental Protection Agency. Office of Air Quality Planning and Standards. *Air Trends*. Research Triangle Park, NC. Compiled annually, and available at http://www.epa.gov/airtrends/.

[2] Martineau, Robert J., Jr. & Novello, David P. (eds.) (2004). *The Clean Air Act Handbook*. 2nd edition. Chicago: American Bar Association. 728 p.

Table 3. Major U.S. Code Sections of the Clean Air Act (Codified Generally as 42 U.S.C. 7401-7671).

42 U.S.C.	Section Title	Clean Air Act, as amended
Subchapter I -	Programs and Activities	
Part A -	Air Quality Emissions and Limitations	
7401	Findings, purpose	Sec. 101
7402	Cooperative activities	Sec. 102
7403	Research, investigation, training	Sec. 103
7404	Research relating to fuels and vehicles	Sec. 104
7405	Grants for air pollution planning and control programs	Sec. 105
7406	Interstate air quality agencies; program cost limitations	Sec. 106
7407	Air quality control regions	Sec. 107

Table 3. (Continued)

42 U.S.C.	Section Title	Clean Air Act, as amended
7408	Air quality criteria and control techniques	Sec. 108
7409	National primary and secondary air quality standards	Sec. 109
7410	SIPs for national primary and secondary air quality standards	Sec. 110
7411	Standards of performance for new stationary sources	Sec. 111
7412	Hazardous air pollutants	Sec. 112
7413	Federal enforcement	Sec. 113
7414	Recordkeeping, inspections, monitoring, and entry	Sec. 114
7415	International air pollution	Sec. 115
7416	Retention of state authority	Sec. 116
7417	Advisory committees	Sec. 117
7418	Control of pollution from federal facilities	Sec. 118
7419	Primary nonferrous smelter orders	Sec. 119
7420	Noncompliance penalty	Sec. 120
7421	Consultation	Sec. 121
7422	Listing of certain unregulated pollutants	Sec. 122
7423	Stack heights	Sec. 123
7424	Assurance of adequacy of state plans	Sec. 124
7425	Measures to prevent economic disruption/unemployment	Sec. 125
7426	Interstate pollution abatement	Sec. 126
7427	Public notification	Sec. 127
7428	State boards	Sec. 128
7429	Solid waste combustion	Sec. 129
7430	Emission factors	Sec. 130
7431	Land use authority	Sec. 131
Part B - Ozone Protection (repealed—new provisions related to stratospheric ozone protection are found at 42 U.S.C. 7671 et seq., below)		
Part C - Prevention of Significant Deterioration of Air Quality		
Subpart I -	Clean Air	
7470	Congressional declaration of purpose	Sec. 160
7471	Plan requirements	Sec. 161
7472	Initial classifications	Sec. 162
7473	Increments and ceilings	Sec. 163
7474	Area redesignation	Sec. 164
7475	Preconstruction requirements	Sec. 165
7476	Other pollutants	Sec. 166
7477	Enforcement	Sec. 167
7478	Period before plan approval	Sec. 168
7479	Definitions	Sec. 169
Subpart II -	Visibility Protection	
7491	Visibility protection for federal class I areas	Sec. 169A
7492	Visibility	Sec. 169B

Table 3. (Continued)

42 U.S.C.	Section Title	Clean Air Act, as amended
Part D -	Plan Requirements for Nonattainment Areas	
Subpart 1 -	Nonattainment Areas in General	
7501	Definitions	Sec. 171
7502	Nonattainment plan provisions in general	Sec. 172
7503	Permit requirements	Sec. 173
7504	Planning procedures	Sec. 174
7505	Environmental Protection Agency grants	Sec. 175
7505a	Maintenance plans	Sec. 175A
7506	Limitations on certain federal assistance	Sec. 176
7506a	Interstate transport commissions	Sec. 176A
7507	New motor vehicle emission standards in nonattainment areas	Sec. 177
7508	Guidance documents	Sec. 178
7509	Sanctions and consequences of failure to attain	Sec. 179
7509a	International border areas	Sec. 179B
Subpart 2 -	Additional Provisions for Ozone Nonattainment Areas	
7511	Classifications and attainment dates	Sec. 181
7511a	Plan submissions and requirements	Sec. 182
7511b	Federal ozone measures	Sec. 183
7511c	Control of interstate ozone air pollution	Sec. 184
7511d	Enforcement for Severe and Extreme ozone nonattainment areas for failure to attain	Sec. 185
7511e	Transitional areas	Sec. 185A
7511f	NO_x and VOC study	Sec. 185B
Subpart 3 -	Additional Provisions for Carbon Monoxide Nonattainment Areas	
7512	Classification and attainment dates	Sec. 186
7512a	Plan submissions and requirements	Sec. 187
Subpart 4 -	Additional Provisions for Particulate Matter Nonattainment Areas	
7513-	Classifications and attainment dates	Sec. 188
7513a	Plan provisions and schedules for plan submissions	Sec. 189
7513b	Issuance of RACM and BACM guidance	Sec. 190
Subpart 5 -	Additional Provisions for Areas Designated Nonattainment for Sulfur Oxides, Nitrogen Dioxide, or Lead	
7514	Plan submission deadlines	Sec. 191
7514a	Attainment dates	Sec. 192
Subpart 6 -	Savings Provisions	
7515	General savings clause	Sec. 193

Table 3. (Continued)

42 U.S.C.	Section Title	Clean Air Act, as amended
Subchapter II -	Emission Standards for Moving Sources	
Part A -	Motor Vehicle Emission and Fuel Standards	
7521	Emission standards for new motor vehicles or engines	Sec. 202
7522	Prohibited acts	Sec. 203
7523	Actions to restrain violations	Sec. 204
7524	Civil penalties	Sec. 205
7525	Motor vehicle and engines testing and certification	Sec. 206
7541	Compliance by vehicles and engines in actual use	Sec. 207
7542	Information collection	Sec. 208
7543	State standards	Sec. 209
7544	State grants	Sec. 210
7545	Regulation of fuels	Sec. 211
7547	Nonroad engines and vehicles	Sec. 213
7548	Study of particulate emissions from motor vehicles	Sec. 214
7549	High altitude performance adjustments	Sec. 215
7550	Definitions	Sec. 216
7551	Study and report on fuel consumption of CAAA of 1977	Sec. 203
7552	Motor vehicle compliance program fees	Sec. 217
7553	Prohibition on production of engines requiring leaded gasoline	Sec. 218
7554	Urban bus standards	Sec. 219
Part B -	Aircraft Emissions Standards	
7571	Establishment of standards	Sec. 231
7572	Enforcement of standards	Sec. 232
7573	State standards and controls	Sec. 233
7574	Definitions	Sec. 234
Part C -	Clean Fuel Vehicles	
7581	Definitions	Sec. 241
7582	Requirements applicable to clean-fuel vehicles	Sec. 242
7583	Standards for light-duty clean-fuel vehicles	Sec. 243
7584	Administration and enforcement as per California standards	Sec. 244
7585	Standards for heavy-duty clean-fuel vehicles	Sec. 245
7586	Centrally fueled fleets	Sec. 246
7587	Vehicle conversions	Sec. 247
7588	Federal agency fleets	Sec. 248
7589	California pilot test program	Sec. 249
7590	General provisions	Sec. 250
Subchapter III -	General Provisions	
7601	Administration	Sec. 301
7602	Definitions	Sec. 302
7603	Emergency powers	Sec. 303

Table 3. (Continued)

42 U.S.C.	Section Title	Clean Air Act, as amended
7604	Citizen suits	Sec. 304
7605	Representation in litigation	Sec. 305
7606	Federal procurement	Sec. 306
7607	Administrative proceedings and judicial review	Sec. 307
7608	Mandatory licensing	Sec. 308
7609	Policy review	Sec. 309
7610	Other authority	Sec. 310
7611	Records and audits	Sec. 311
7612	Economic impact analyses	Sec. 312
7614	Labor standards	Sec. 314
7615	Separability	Sec. 315
7616	Sewage treatment plants	Sec. 316
7617	Economic impact assessment	Sec. 317
7619	Air quality monitoring	Sec. 319
7620	Standardized air quality modeling	Sec. 320
7621	Employment effects	Sec. 321
7622	Employee protection	Sec. 322
7624	Cost of vapor recovery equipment	Sec. 323
7625	Vapor recovery for small business marketers of petroleum products	Sec. 324
7625-1	Exemptions for certain territories	Sec. 325
7625a	Statutory construction	Sec. 326
7626	Authorization of appropriations	Sec. 327
7627	Air pollution from Outer Continental Shelf activities	Sec. 328
Subchapter IV-A	Acid Deposition Control	
7651	Findings and purposes	Sec. 401
7651a	Definitions	Sec. 402
7651b	Sulfur dioxide allowance program for existing and new units	Sec. 403
7651c	Phase I sulfur dioxide requirements	Sec. 404
7651d	Phase II sulfur dioxide requirements	Sec. 405
7651e	Allowances for states with emissions rates at or below 0.80 lbs./mmBtu	Sec. 406
7651f	Nitrogen oxides emission reduction program	Sec. 407
7651g	Permits and compliance plans	Sec. 408
7651h	Repowered sources	Sec. 409
7651i	Election for additional sources	Sec. 410
7651j	Excess emissions penalty	Sec. 411
7651k	Monitoring, reporting, and recordkeeping requirements	Sec. 412
7651l	General compliance with other provisions	Sec. 413
7651m	Enforcement	Sec. 414

Table 3. (Continued)

42 U.S.C.	Section Title	Clean Air Act, as amended
7651n	Clean coal technology regulatory incentives	Sec. 415
7651o	Contingency guarantee, auctions, reserve	Sec. 416
Subchapter V -	Permits	
7661	Definitions	Sec. 501
7661a	Permit programs	Sec. 502
7661b	Permit applications	Sec. 503
7661c	Permit requirements and conditions	Sec. 504
7661d	Notification to administrator and contiguous states	Sec. 505
7661e	Other authorities	Sec. 506
7661f	Small business stationary source technical and environmental compliance assistance program	Sec. 507
Subchapter VI -	Stratospheric Ozone Protection	
7671	Definitions	Sec. 601
7671a	Listing of class I and class II substances	Sec. 602
7671b	Monitoring and reporting requirements	Sec. 603
7671c	Phase-out of production and consumption of class I substances	Sec. 604
7671d	Phase-out of production and consumption of class II substances	Sec. 605
7671e	Accelerated schedule	Sec. 606
7671f	Exchange authority	Sec. 607
7671g	National recycling and emission reduction program	Sec. 608
7671h	Servicing of motor vehicle air conditioners	Sec. 609
7671i	Nonessential products containing chlorofluorocarbons	Sec. 610
7671j	Labeling	Sec. 611
7671k	Safe alternatives policy	Sec. 612
7671l	Federal procurement	Sec. 613
7671m	Relationship to other laws	Sec. 614
7671n	Authority of Administrator	Sec. 615
7671o	Transfers among parties to Montreal Protocol	Sec. 616
7671p	International cooperation	Sec. 617
7671q	Miscellaneous provisions	Sec. 618
[29 U.S.C. 655]	Chemical Process Safety Management	Sec. 304 of CAA of 1990
[29 U.S.C. 1662e]	Clean Air Employment Transition Assistance	Sec.1101 of CAA of 1990

Note: This table shows only the major U.S. Code sections. For more detail and to determine when a section was added, the reader should consult the official printed version of the U.S. Code.

End Notes

[1] Unlike the other NAAQS pollutants, ozone is not directly emitted, but rather is formed in the atmosphere by the interaction of volatile organic compounds (VOCs) and nitrogen oxides (NOx) in the presence of sunlight. The control of ozone is, thus, based on regulating emissions of VOCs and NOx.

[2] EPA modified the ozone standard, specified in the statute as 0.12 parts per million (ppm) averaged over a 1-hour period, to 0.08 ppm averaged over an 8-hour period, through regulations promulgated in July 1997. In April 2004, the agency promulgated an implementation rule for the new 8-hour standard. Under this rule, the 1-hour standard was revoked as of June 15, 2005, and areas that had not yet attained it were converted to new classifications depending on their 8-hour concentration of ozone. As a result of court challenges, the ramifications of this conversion to the 8-hour standard are still unfolding, but in general the former 1-hour nonattainment areas remain subject to the controls specified for their 1-hour category. New nonattainment areas that did not exceed the 1-hour standard, but do violate the 8-hour standard, in general are subject to more flexible controls under Subpart 1 (Sections 171-179B) of the act. The standard was revised again in March 2008, to 0.075 ppm averaged over 8 hours, but EPA subsequently agreed to reconsider the 2008 standard, a task it expects to complete in 2011. Nonattainment areas for the new standard would be designated following its promulgation.

[3] This section of the report was written by Linda-Jo Schierow, Specialist in Environmental Policy.

[4] The 1990 amendments specified 189 pollutants, but P.L. 102-187, enacted on December 4, 1991, deleted hydrogen sulfide from the list of toxic pollutants, leaving only 188. On December 19, 2005, EPA removed methyl ethyl ketone (MEK) from the list of toxic air pollutants. The total number of listed air toxics is now 187.

[5] This section of the report was written by Larry B. Parker, Specialist in Energy and Environmental Policy.

[6] This section of the report was written by Larry B. Parker, Specialist in Energy and Environmental Policy and James E. McCarthy, Specialist in Environmental Policy.

[7] This section of the report was written by Larry B. Parker, Specialist in Energy and Environmental Policy.

[8] This section of the report was written by Claudia Copeland, Specialist in Environmental Policy.

[9] This section of the report was written by Larry B. Parker, Specialist in Energy and Environmental Policy.

INDEX

C

F

G

H

I

J

K

L

M

T

U

V

W

Y

Z